ALGORITHMS OF
RESISTANCE

ALGORITHMS OF RESISTANCE

THE EVERYDAY FIGHT AGAINST PLATFORM POWER

TIZIANO BONINI AND EMILIANO TRERÉ

THE MIT PRESS   CAMBRIDGE, MASSACHUSETTS   LONDON, ENGLAND

The MIT Press would like to thank the anonymous peer reviewers who provided
comments on drafts of this book. The generous work of academic experts is essential for establishing the authority and quality of our publications. We acknowledge
with gratitude the contributions of these otherwise uncredited readers.

This book was set in Stone Serif and Avenir by Westchester Publishing Services.
Printed and bound in the United States of America.

Library of Congress Cataloging-in-Publication Data

Names: Bonini, Tiziano, author. | Treré, Emiliano, author.
Title: Algorithms of resistance : the everyday fight against platform power /
  Tiziano Bonini and Emiliano Treré.
Description: Cambridge, Massachusetts : The MIT Press, [2024] | Includes
  bibliographical references and index.
Identifiers: LCCN 2023018017 (print) | LCCN 2023018018 (ebook) |
  ISBN 9780262547420 (paperback) | ISBN 9780262377492 (epub) |
  ISBN 9780262377485 (pdf)
Subjects: LCSH: Internet—Social aspects. | Online algorithms—Social
  aspects. | Online algorithms—Moral and ethical aspects. | Online
  manipulation. | Gig economy. | Social media and society.
Classification: LCC HM851 .B6747 2024 (print) | LCC HM851 (ebook) |
  DDC 302.23/1—dc23/eng/20230628
LC record available at https://lccn.loc.gov/2023018017
LC ebook record available at https://lccn.loc.gov/2023018018

10 9 8 7 6 5 4 3 2 1

This book is dedicated to all the workers, activists, and users who shared with us their tactics of resistance to the power of digital platforms and their algorithms. You are the cracks through which the light gets in.

One tail, One train.
—Andre Leyton, *Snowpiercer*, 2020

CONTENTS

ACKNOWLEDGMENTS xi

INTRODUCTION 1

1 LIVING WITH ALGORITHMS: POWER, AGENCY, RESISTANCE 13

2 THE MORAL ECONOMY OF ALGORITHMIC AGENCY 29

3 GAMING THE BOSS 59

4 GAMING CULTURE 107

5 GAMING POLITICS 131

6 FRONTIERS OF RESISTANCE IN THE AUTOMATED SOCIETY 155

APPENDIX: RESEARCH METHODS 179
NOTES 189
INDEX 227

ACKNOWLEDGMENTS

We started talking about the idea of this book in 2018. It took us a few years to generate the data, and at least two years more to write it. In between, there was a global pandemic and several personal crises. If we made it to the end, it is certainly not just because of us. Every book is a sociotechnical apparatus, an ensemble of human and nonhuman actors. We are the ones who put our names on the cover, but behind that, there is a network that has made this book possible. First of all, we would like to give our heartfelt thanks to William Uricchio, who, despite the fact that we met only once at the 2019 MIT Media in Transition 10 conference in Boston, responded with great enthusiasm to our first email asking him for his opinion on our book proposal. As our first reader, he provided highly valuable comments to review the project before submission, and his support was thus key to the success of this book. Next, we would like to thank our editor, Noah Springer, who gave us enthusiasm and confidence from the first time we met. Without him, the book would not be here in this form.

We would also like to thank the researchers who helped us generate the data and who embarked on the Algorithmic Resistance project with us during the global pandemic: Zizheng Yu, Swati Singh, Daniele Cargnelutti, and Francisco Javier López Ferrández are all incredibly gifted early career scholars, and the dialogue with them has enabled us to develop our ideas further. Despite the increasing precarization of academic work and of all cultural industries, we hope that they will be able to continue doing what they do so well and shine as leading academics.

We would like to thank Nadim and Bruna, the Florentine couriers who provided us with unvaluable insights on their everyday life as gig workers and accepted to read and review the findings described in chapter 3. We also thank Giulia Druetta, a Turin lawyer that, through her lawsuits against food delivery platforms, helped defend the rights of platform workers. Thank you for providing valuable insights into how these platforms work.

Special thanks go to Francesca Murtula, an MA student supervised by Tiziano at the University of Siena, who worked with us both as a research assistant on this book and as the coauthor of related articles. Emiliano would like to thank his academic home, the School of Journalism, Media, and Culture at Cardiff University for providing funding to hire Thomas Davis as a research assistant supporting the project. His work has been invaluable in providing us with rich examples of algorithmic agency and resistance from all over the world. We are immensely grateful to all the brilliant research assistants who contributed to the research that informs our book. There are also many colleagues whom we would like to thank for providing comments, support, and friendship during the fieldwork and the writing of the book. In particular, we are grateful to José van Dijck, William Uricchio, Ignacio Siles, Robert Prey, Adam Arvidsson, Alessandro Delfanti, Guido Smorto, Riccardo Pronzato, Gabriel Pereira, and Luke Heemsbergen for their helpful comments.

Tiziano would like to thank his close circle of friends and colleagues who have supported him over the years, starting with Adam Arvidsson, Bertram Niessen, Davide Sparti, Tarcisio Lancioni, Alessandro Gandini, Alessandro Caliandro, Carolina Bandinelli, Alberto Cossu, Alessandro Delfanti, Paolo Magaudda, Ignacio "Nacho" Gallego Pérez, Robert Prey, Belén Monclús, Luis Albornoz, and all the ECREA Radio research community. My intellectual work is heavily indebted to the long conversations we have had over the years.

Emiliano is deeply indebted to the people whose help, passion, and support have been invaluable during the writing of this book, including Ana Müller ("las palabras nunca alcanzan cuando lo que hay que decir desborda el alma"), Stefania Milan, Silvia Masiero, Veronica Barassi, Gabriela Sued, Dorismilda Flores, Rossana Reguillo, Elisenda Ardèvol, Francesca Comunello, Eva Campos, Ángel Barbas, Simone Natale, Alejandro Barranquero, Jose Candón, Christian Schwarzenegger, Antoni Roig, Carlos Scolari, Guillermo López García, and the Data Justice Lab (Lina Dencik, Joanna Redden, and Arne Hintz).

Finally, we would like to thank our families.

Tiziano would like to thank his wife, Ilaria, for her constant love, trust, and active listening. She always knows how to give the right weight to the events of life. For Tiziano, the writing of this book coincided with the long wait for the arrival of Lea, his first daughter. Lea was eventually born just as we were finishing revising the drafts of the book. We wish her a long life, full of love and opportunities to express her agency.

Emiliano would like to express his deepest gratitude, appreciation, and love to Leo and Aloia for letting me be who I am through the words that I write.

A final dedication goes to the Italian public university system, which made us study (almost) for free.

INTRODUCTION

STEFANO

Stefano[1] is forty-three years old. He has a daughter and a partner, works as a photographer and video maker, and lives in Livorno, a seaport city in Tuscany, Italy. For many years, he has been working as an art photographer and, from time to time, for advertising. During the first COVID-19 lockdown, he lost many job opportunities and decided to download the Deliveroo app on his smartphone to start making home deliveries. He fixed an old bike he had in his garage and started riding. The first few days were very good, Stefano recalls. In a week, he had already earned almost 300 euros, and the work gave him a strange sense of freedom: making easy money, without talking to anyone, without having to submit to some annoying boss. Stefano is a freelancer in life, and he is used to working alone, so he really appreciated the freedom that Deliveroo seemed to afford him.

Flash forward a month later. Stefano tells us that he is no longer so happy to use Deliveroo. On the contrary, he is a bit worried because sometimes he finds himself stuck on the Deliveroo app to check if there are any free shifts that can be assigned to him. He feels strangely dependent on the app, and he doesn't like that. His feeling of dependence is similar to that experienced by other gig workers, such as Uber drivers: "The experience of this job feels like an addiction because it feels nice at first, then it's really bad."[2]

In the meantime, the available shifts have decreased, and Stefano doesn't earn as much money as before. One Saturday night, he forgot to cancel his reservation for a shift that he had been assigned because he had to attend the opening of his new photo studio. From that moment on, he realized that the app was assigning him shifts only in time slots where there were few deliveries to make, so he earned less. Chatting with other couriers like him in the WhatsApp group created by Livorno's delivery gig workers, he discovered that his forgetfulness costed him dearly because "the algorithm lowered [his] statistics," and he is no longer considered 100 percent reliable, but only 98 percent.

By participating in the conversations within the WhatsApp group, he learned several things about how Deliveroo's algorithm works, or rather, how the couriers of Livorno imagine that the algorithm works. A Sardinian guy taught him some tricks to cancel an assignment without losing points. "Without the support of this WhatsApp group," Stefano recalls, "I would have already quit the job."

Thanks to Stefano, whom we met through a mutual friend, we get in touch with other couriers from Livorno, and they put us in contact with others from Florence, Naples, and Milan. While some of them have been doing this job for only a few months, others have been active for more than two years. We hang out with some of them for the whole summer of 2020, asking them to keep an audio diary and to send us WhatsApp audio notes, in which they reflect on their tactics to get better shifts from the app and earn more money. Everyone told us the same thing: at first, they were enthusiastic about the app and the speed with which they could make money and round out their income, but at a certain point, they started earning less and feeling more dependent on the app. While at first they didn't know anything about the algorithm that governs the app, today they say that they are quite experienced, have a pretty good idea of how it works, and have developed tricks to "cheat the app."

ALGORITHMS OF RESISTANCE

What gig workers call "tricks," we call manifestations of *tactical algorithmic agency,* or the ability of the people to actively shape the outcome of algorithmic computation for their own benefit. What are these tactics that we talk about in this book? How powerful are they? And what does it mean to *resist* algorithmic power?

The title of this book evokes the possibility that algorithms may also be employed to resist the power of those who programmed them. Ours is not a cyberpunk fantasy, but rather an awareness grounded in the practices that we have observed during years of research. The valuable work of scholars such as the American sociologist Safiya Umoja Noble has given us a deep understanding of how algorithms can be sources of discrimination and oppression. With our work, however, we want to show another side of the issue—namely, that algorithms, as well as producing oppression, can also be appropriated by users to resist the power of technology companies. They can be both: algorithms of oppression and algorithms of resistance.

We will venture into uncharted algorithmic territories with the aim of discovering and mapping all those forms of agency and those practices of resistance and resilience that users of digital platforms put in place to survive in the *chaos* of the platform society. This set of practices, whose existence we found in different cities around the world and in multiple realms of social life—from the mundane to the political—constitutes our map of *algorithmic agency*.

Our bodies and actions are continuously calculated and transformed into flows of data that feed the platforms' algorithms. We are constantly subjected to a process of stripping and extracting biometric, biographical, and demographic data that in the era of big data has been called *datafication*,[3] but which, as Colin Koopman has illustrated, has a long history behind it.[4] Some media scholars, such as Nick Couldry and Ulises Mejias, call this process *data colonialism*,[5] while Shoshana Zuboff attributes to platforms an *instrumentarian power* capable not only of extracting from our simple being in the world a *behavioral surplus*, but also of automating our choices.[6] Online platforms—whether American, Chinese, Russian, or to a lesser extent European—have acquired an enormous power, which scholars of the political economy of media call *platform power*,[7] and it conditions the emerging platform society. We users receive unquestionable benefits from using these platforms, but at the same time, we are caught within an incredibly asymmetrical power relationship. This means, among other things, that users are not endowed with the same computational power that the platforms have.

Of course, we are not all equally powerless and vulnerable. Some are more so than others. As Virginia Eubanks has shown,[8] the poor are more exposed to the discrimination generated by algorithms and the power of platforms. Moreover, this power risks discriminating not only against the poor, but also against ethnic and linguistic minorities, indigenous peoples, young people, and women. Algorithmic discrimination can also take intersectional forms, affecting, for example, young single women and ethnic minority families belonging to the working class, as happened in the Netherlands and Australia.[9]

Yet we will see how people enact different tactics to reduce this asymmetry, as well as how they are able to attribute new meanings to the algorithms they use, transforming them into effective tools for pursuing their own political, economic, cultural, or social agendas. Having less power than digital platforms, however, does not automatically mean being on the side of the "good guys." User agency, as we shall see, can also generate actions that some of us might consider reprehensible or criminal.

AGENCY ACROSS GIG WORK, CULTURE, AND POLITICS

Among the many categories of users who confront the power of platforms every day, we chose to deal with three of them: those who use platforms to work, to produce and consume cultural objects, and to engage in political activity. Gig workers (work), artists, musicians, fandom, and content creators (culture), and social movements and political parties (politics) will be the main protagonists of this book. What do these subjects have in common? They are all, in some way, digital laborers. They all perform digital work. Yet the digital labor performed every day by gig workers, cultural producers, and political activists is not all the same. As the Italian digital labor scholar Alessandro Gandini has rightly noted, attributing the label of "digital labor" to all the activities performed online or mediated by digital platforms risks turning the term itself into an "empty signifier."[10] Gandini therefore proposes to distinguish between the *free labor* exercised by social media users acting as unpaid content producers and the *platform labor* exercised by underpaid (gig) workers. While cultural producers' and political activists' labor is exploited by "content media" companies, gig workers' labor is exploited by "content*less* media" companies[11] such as digital work platforms (Upwork, Uber, Airbnb etc.). The free labor discussed by critical scholars like Tiziana Terranova, Trebor Scholz, and Christian Fuchs[12] represents the digital version of the exploitation of the audience's ability to "pay attention" operated by broadcast media.[13] Instead, according to Gandini, the subjects involved in platform labor "are not viewers or users whose leisure activity is exploited but actual workers who willingly subject themselves to the execution of activities commissioned by a customer/client through a digital platform, which effectively acts as a 'shadow' or 'pseudo' employer."[14] In this latter case, exploitation shifts from the "'consumption work' of audiences and the datafication of the 'productive leisure' of social media users"[15] to digitally mediated physical work increasingly surveilled and datafied.[16]

However, whether the digital work performed by this new multitude of subjects is content-oriented or contentless, or whether it is unpaid or indecently underpaid, it always runs up against the same algorithmic logic that underlies both the selection of labor and media content. In this book, we will show that both free laborers and digital workers, in the face of the disproportionate computational power wielded by tech companies, painstakingly manage to improve their visibility and their working conditions, organize forms of collective action, and build solidarity bonds. They can all exercise varying degrees of algorithmic agency, regardless of their situated labor status.

We also chose these three domains—gig work, cultural industries, and politics—because they represent three important moments of our everyday lives and we want to show how the ability to exercise agency cuts across various spheres of life and is common to various online platforms. Our research is therefore limited to these three meaningful domains of everyday life, but there are many others where algorithms play an increasingly significant role: in educational, health-care, and financial institutions, in welfare state systems, and in public administrations, to name just a few. In these domains too, there is a need for research into the kind of agency available to people and the forms of resistance emerging from below.

Through the account of what people do with algorithms in gig work, culture, and politics, we want to show how ordinary people have an unsuspected ability to invent practices and adopt tactics to evade (even if only temporarily) the constraints of algorithmic power, especially when they are able to organize and act collectively.

Our focus, therefore, is the agency of users in relation to the power of platforms, and one of our aims is to develop a conceptual framework to study the agency that people have in a platformized environment. So this book can also be described as an empirical and theoretical exploration of human agency in relation to the power of platforms. This exploration will lead us, as we shall see, to break down the concept of agency and recompose it in a multidimensional way. We will propose that users' agency in relation to algorithms—which we will call *algorithmic agency*—can be expressed in four ideal-typical models, and we will show how these four manifestations of algorithmic agency can be found in all the domains of work, culture, and politics.

The risk of such work is falling into the usual trap of overestimating user agency, as media audiences' agency has been overemphasized in the past. Our interest in the agency of platform users is not aimed at disavowing or minimizing the power of platforms. On the contrary, we want to provide a more complex narrative, in which power relations between users and platforms are never frictionless and never taken for granted. Underlying all the analyses that have overstated agency in the past is, as Ien Ang pointed out back in the 1990s, a key conceptual confusion: improperly, these analyses "cheerfully equate [the active] with the powerful."[17] This is a mistake that we avoid in our book: showing that people are active and that, despite everything, they can exercise different kinds of algorithmic agency does not mean that they also have "power." As Ien Ang continued in the article just cited, "We must not lose sight of the *marginality* of this power."[18]

The users described here have developed a repertoire of actions that testifies to the existence of a certain degree of agency at their disposal. The range of action of this agency, however, is highly variable and constrained by the structural limits imposed by the technological affordances of the platforms and their terms of service (ToS).

We are therefore interested in showing that power relations, however asymmetrical they may be, are dynamic, contingent, socially constructed, and constantly renegotiated.

OUTLINE OF THE BOOK

In the last decade, a growing body of studies have converged their focus on exploring the social, material, and political dimensions of the technological infrastructures that underlie the global flows of information and communication. So far, however, there have been few studies on how the power of these infrastructures has been accepted, negotiated, and incorporated into people's daily lives.

In chapter 1, we analyze three foundational keywords of the whole book: platform power, human agency, and algorithmic resistance. We critically review the ongoing debate on agency and algorithmic power, highlighting that contemporary studies on digital platforms, algorithmic media, and data colonialism often provide rather monolithic accounts of algorithmic power and tend to leave forms of agency and resistance unattended. We then dive into the ongoing debate on the definition of agency in relation to the power of algorithms. Building on Giddens's structuration theory[19] and his reflections on the duality of structure, we argue that human agency and algorithmic infrastructures mutually shape each other, and we frame the relationship between users and algorithmic infrastructures as a symbiotic one. We advance the concept of *algorithmic agency*, situating it within this symbiotic approach. We subsequently move on to our third keyword, *resistance*. We argue that there is neither a clear distinction between agency and resistance nor a perfect superposition. Rather, we propose that the agency manifestations that we describe here move along a *continuum* that goes from forms of agency that openly resist platform power and other forms of agency that have no intention of questioning or challenging platform power. To make this distinction, however, we first define what we mean by resistance, drawing on the works of Jocelyn Hollander and Rachel Heinwhoner,[20] James Scott,[21] and Michel de Certeau.[22] Further, we distinguish between three forms of algorithmic resistance, explaining what we mean by resistance to and through algorithms.

Chapter 2 introduces and discusses the key theoretical framework of the book. First, it proposes to understand the forms of algorithmic agency and resistance through the concept of the moral economy.[23] The chapter articulates why the moral economy is helpful to capture the daily relationship that people have with algorithmic infrastructures by situating the various forms of algorithmic agency along a continuum shaped by moral values that sees, at its two extremes, two competing moral economies: the user moral economy and the platform moral economy. It demonstrates that this conceptual lens allows us to defuse the rhetoric inscribed in the too-simplistic "gaming versus optimization" distinction, showing how the assignment of a morally negative (gaming) or positive (optimization) value to a certain practice depends on the type of moral economy embraced by the subject. Recognizing that the "moral dimension" is not enough, the discussion introduces a further dimension, bringing into the equation the type of power held by those who enact algorithmic resistance. To the poles represented by the two opposing moral economies of the users and the platforms, located along a horizontal axis, it adds the strategic versus tactical poles located along the vertical axis. These two dimensions—the moral and the tactical/strategic—make up the theoretical framework that lies at the heart of this book. This framework accounts for four ideal-typical forms of agency available to individual people and institutions in their engagements with algorithmic power.

Chapters 3, 4, and 5 represent the core of the field research conducted for this book. Each chapter deals with a different field—gig work, culture, and politics—in which we investigated the emergence of algorithmic agency. The chapters are titled "Gaming the Boss," "Gaming Culture," and "Gaming Politics." In chapter 2, we will explain why we believe that "gaming" is not the proper frame for understanding the meaning of all the manifestations of algorithmic agency that the reader will encounter in the book. Yet we decided to keep the word "gaming" in the title of these three central chapters for two reasons. The first, more mundane one, is that it is an immediately understandable and evocative term. The second, however, is related to its multiple layers of meaning: while technology companies load this term with negative meaning, users take it as positive: we noticed that "gaming the system" can be a lot of fun and can generate a sense of pride among users. Therefore, when we use this term in the title of these chapters, we are referring to this double meaning. Our choice to use the term "gaming" in relation to "work, culture, and politics" thus stands for the point of view not only of technology companies, but also of users, and it helps us situate work, culture, and politics as battlegrounds in which companies and users constantly negotiate the meaning of their practices directed at optimizing their profits (companies) or their visibility (users).

Chapter 3 zooms in on the algorithmic agency and resistance emerging among gig workers. We account for different practices of algorithmic agency and resistance in the realm of the gig economy: from "surge clubs" among Uber and Lyft drivers to individual and collective tactics and strategies created by workers of online food delivery platforms like Deliveroo. We draw from our fieldwork and interviews with couriers and drivers of the online food delivery platforms in India, China, Mexico, Italy, and Spain and case studies from the public press in both the Global North and Global South. The data generated from the field show that forms of resistance to the work of the algorithms are rational practices driven by a different moral economy than the one coded into the algorithms governing online food delivery platforms.

Chapter 4 focuses on the manifestation of algorithmic agency and resistance in cultural industries. First, we analyze how the rise of platforms is changing traditional cultural industries while reshaping cultural creation, distribution, and consumption. We critically review the emerging research around the process called by Nieborg and Poell "platformization of cultural industry," and then we summarize the contemporary conditions of platformized cultural work. We frame cultural work as an increasingly precarious activity based on visibility labor and argue that visibility is more central than ever in the valuation of cultural work. On the one hand, online platforms developed a technical infrastructure that could calculate, datafy, and commodify visibility; on the other hand, wherever visibility is at stake, we find individual and collective practices that attempt to artificially manipulate and reappropriate it. Visibility is thus the battleground where platforms and cultural workers confront each other. In the second half of the chapter, we show a practical example of the efforts made by cultural workers to "game" visibility and we focus on the case study of engagement groups (*pods*) on Instagram. We rely on an eight-month-long digital ethnography to understand the processes of meaning-making that take place within them. We finally propose to frame the activity of these groups as a manifestation of collective agency and show how these groups represent forms—albeit fragile and temporary—of resistance to the power and moral code of online platforms.

Chapter 5 examines the manifestation of algorithmic agency and resistance in the realm of politics, proposing the notion of *algorithmic politics*. We locate it within the broader scenario of data politics, drawing a distinction between an *institutional/strategic* type of algorithmic politics and a *contentious/tactical* one (which we identify with *algorithmic activism*). In the first part, relying on case studies from Europe to Latin America, from the US to North Africa and Asia, the chapter illustrates strategies of institutional/strategic algorithmic politics. It shows how algorithms have been used to manipulate public opinion, spread propaganda, create an illusion of popularity, and undermine digital dissent. Then, we illuminate how algorithmically mediated environments radically restructure the dynamics of activism, collective action, and the repertoires of contention of social movements. We then move on to unravel the contemporary technopolitical battlefield, characterized by an incessant back-and-forth between strategies and tactics, fleshing out a taxonomy of three types of political engagement with algorithms (*amplification*, *evasion*, and *hijacking*). Drawing on several global case studies, as well as on in-depth interviews carried out with activists in Europe and Latin America, we show that social media algorithms have material impacts on the emergence and dynamics of social movements and the diffusion of protest. In the conclusion, we discuss the moral economy of algorithmic activism, interrogating the connections and differences with the concept of hashtag activism and reflecting on the fact that algorithms have been equally appropriated by both conservative and progressive social movements (what we call the "agnosticism" of algorithmic activism).

In chapter 6, we summarize the key contributions of the book and reflect on its conceptual journey, establishing further connections with broader debates on automation, artificial intelligence (AI), algorithmic power, datafication, platform capitalism, and resistance. We stress the relevance of the key concepts put forward in the book: the multidimensionality of algorithmic agency, the existence of different moral economies within the platform society, and the importance of everyday forms of algorithmic resistance for the construction of more structured models of resistance to platform power. This final chapter aims to provide a response to the pessimistic narratives of the future that awaits us, such as the nihilistic ones, à la Mark Fisher,[24] which sound desperate because scholars like him believe that there is no alternative to this model of production and development. The answer to this nihilism does not lie in a romantic vision of human agency, but rather in a Gramscian vision of the relationship between citizen agency and the structure of the platform society: pessimism of the intellect and optimism of the will.[25] This mix of pessimism and optimism is not just an exercise in wishful thinking; on the contrary, it is grounded in our fieldwork. As much as platforms are expanding their power, people are not accepting to passively experience it and are organizing themselves to strike back in various fields. Hence, there is not yet a completely dominant narrative around the power of algorithms in society.

All the domains that we have explored through our fieldwork (the gig economy, cultural consumption, politics, and activism) are characterized by users who organize themselves into online groups—on WhatsApp, Telegram, Signal, or other platforms—to orchestrate collective actions aimed at affecting algorithms. Tech companies and the media label these practices as attempts of "gaming the algorithm" and have often described them as immoral or illegal.[26] We will explain why the "gaming" frame fails to account for the richness of these practices. Gaming represents only the public mask of these actions, the frame imposed from above by online platforms and the media. We have drilled down to bring to light what lies beneath these gaming efforts. In the case of gig work, this means moving from the public profile of the workers, with smiling faces and compliant attitudes toward their clients, to the private domestic spaces of their homes or the private chats that they share with their colleagues and friends. As we immersed ourselves in the observation of these chats and online groups where their members take off the mask of obedient users of the platforms, we have gradually experienced the emergence of a complex network of mutual aid groups and resistance practices, often temporary or limited in duration, which demonstrate the existence of a power relationship between platforms and users that is much less passive, and much more contradictory, than the one commonly portrayed by the media.

A NOTE ON METHODOLOGY AND POSITIONALITY

The book is the result of various years of research and fieldwork and is informed by a set of data generated through qualitative methods. For a more in-depth look at the research methodology and how we generated the data, please refer to the detailed methodological appendix at the end of this book. We are aware, however, that each researcher generates data from a specific position, and therefore, we believe that it is necessary to disclose some elements of the authors' backgrounds that might have informed the analysis presented in this book. Both of us are white male researchers born in Italy. Both are first-generation academics with a background in media and communication studies, working in institutions located in the Global North. We also share the same background as media and political activists at the turn of the 1990s and early 2000s. Our first language is Italian, but we feel at home speaking Spanish and English as well. Both of us are cisgender.

Despite being born and raised within working-class and provincial lower-middle-class families in the center of Italy, we acknowledge that our class-related experiences are different from those of the interviewees, and our position as researchers living and working in the Global North provides us with privilege that the study participants do not hold. As academics and Western citizens, we have enough cultural, social, and economic capital at our disposal to confront the power of platforms and make informed choices about platform consumption. In brief, we can choose whether, how much, and how to use them, whereas many of our interviewees did not have that choice.

Writing this book was a long adventure for us, and throughout this journey, we constantly reflected on the limits of our gaze and how much our analysis was conditioned by our position, but in any case, we are aware that writing is neither neutral nor innocent. As the American sociologist Howard Becker once noted, "In the greatest variety of subject matter and in work done by all different methods at our disposal, we cannot avoid taking sides, for reasons firmly based in social structure."[27]

Florence–Cardiff

June 2020–December 2022

1

LIVING WITH ALGORITHMS: POWER, AGENCY, RESISTANCE

INTRODUCTION

This chapter critically reviews the ongoing debate on agency and algorithmic power, arguing that contemporary studies on digital platforms often provide rather monolithic accounts of algorithmic power that tend to leave forms of agency and resistance unattended. We conceptualize the relationship between users and algorithmic infrastructures as a symbiotic alliance, and we advance two of the conceptual pillars of this book—namely, *algorithmic agency* and *algorithmic resistance*.

PLATFORM POWER AND ITS LIMITATIONS

In the last decade, a growing body of research in media studies has focused on the social, material, and political dimensions of the technological infrastructures that underlie the global flows of information and communication.[1] The increasing spread of technological platforms controlled by a small group of corporations (the so-called GAFAM—Google, Amazon, Facebook, Apple, and Microsoft;[2] and BAT—Baidu, Alibaba, and Tencent[3]) and the penetration of these companies into a growing number of areas of social, cultural, and economic life[4] have led scholars to focus their attention on the increasing power of these platforms and their impact on society. This rise of platform and infrastructural studies[5] has critically interrogated the social consequences of platform power.[6] As authors such as Ganaele Langlois and Greg Elmer[7] have illustrated, when social media platforms expand to reach a quasi-infrastructural scale, their realm of data capture greatly expands. Digital platforms are now playing an increasingly central role in the sorting, categorizing, and hierarchizing of cultural products and commercial services. The progressively central position acquired by these technological giants in the social, economic, and cultural lives of citizens around the world has prompted many scholars to turn their attention to the consequences and implications of the platformization of everyday life and society. While media scholars in the twentieth century investigated the effects of mass media on society, more recent studies have focused instead on the effects of digital platforms on society. Platform power is increasingly pervasive, opaque, and asymmetrical and is fueled by data. The Harvard professor Shoshana Zuboff[8] has coined the term *instrumentarian power* to capture the specific form of power exercised by the recent mutation of industrial capitalism into surveillance capitalism. This new research trend has represented a key turning point in the advancement of media studies and their neighbouring disciplines.

Parallel to the emergence of platform studies, in the second decade of the twenty-first century, media and Internet scholars began to focus on the digital infrastructures that enable communication, emphasizing the material aspects of the platform society. In a relatively short period, the concept of infrastructure become so fashionable in media and communication studies that we started talking of an "infrastructural turn"[9] in media and Internet research. This conceptual turn has been fundamental to explore the world-making dimensions of media and communication systems that had not been previously sufficiently interrogated. As Tarleton Gillespie et al. have remarked, "in communications and media scholarship, the overwhelming focus has been on texts, the industry that produces them, and the viewers that consume them; the materiality of devices and networks has been consistently overlooked."[10] This renewed attention to the material aspect of technological infrastructures is not, however, a recent "discovery" of media studies, but rather a "continuous low" of this discipline.[11]

Such scholars as the media historian John Durham Peters[12] have contributed to bringing back to center stage a reflection on media as infrastructures rather than texts, but before him, the so-called Canadian school[13] had already highlighted the material properties of the media and their consequences on the political organization of states and empires. Hence, this infrastructural turn has been fundamental to ground the presumed immateriality of cyberspace and digital capitalism and to shed light on the political, economic, and environmental impacts of these infrastructures. We consider these studies to be of extreme academic and public value, and we integrate their key insights and lessons, especially their foregrounding of the systemic causes and structures of oppression that define the effects of algorithms on society.

Recently, other scholars coming from disciplines such as critical data, algorithm, and design studies have shed light on other kinds of negative effects of platform power—namely, the biases that are often reproduced by the proprietary algorithms of these platforms. These accounts have made a great contribution to our understanding of the potentially devastating effects that algorithms can have on society, democracy, and culture.[14] Algorithmic bias and discrimination scholars have contributed to raise awareness among citizens and global civil society about the many perils of an algorithmically governed society. They have illuminated and critically tackled the many problematic assumptions and decisions in relation to race, gender, status, and class and the various forms of oppression and discrimination that are encoded, perpetuated, and exacerbated by algorithmic systems and digital platforms. Parallel to the rise of platform and infrastructure studies and the emerging critical algorithmic studies' literature described here, a "decolonial turn" has emerged in data and technology research.[15] Authors such as Nick Couldry and Ulises Mejias[16] have illustrated the colonial continuities of extraction and exploitation of land, labor, and relations through data systems, theorizing the existence of a new social order in which data relations enact a new form of data colonialism. This new kind of colonialism relies on the exploitation of human beings and the capitalization of life through data, just as historical colonialism appropriated territory and resources and ruled subjects for profit.

Relatedly, the contributions of critical data and algorithm studies are inestimable, as they have focused our attention to multiple forms of discrimination, oppression, and injustice enshrined in algorithmic systems and proposed ways to shape more just and equitable datafied societies.[17]

In connection to these critical strands, the decolonial turn has brought to the fore the urgency to address and excavate forms of extraction, accumulation, exploitation, and injustice that connect contemporary algorithmic media to our colonial past and present.

Yet, in realigning the scholars' attention to how platforms, data centers, software, and algorithms influence society and profoundly reshape media industries, exercising an ever-more-capillary form of power, there is a risk of losing sight of the space still available to people to resist this power. Scholars like Jathan Sadowski[18] and Shoshana Zuboff[19] specifically emphasize the power of digital capitalism and its capacity to strongly determine our lives and automate our taste and consumption decisions. Particularly, Zuboff claims that data collection and the use of predictive algorithms by tech industry corporations represent a means of behavioral modification capable of making human behavior not only completely predictable and manageable but also automated through a "digital order that thrives within things and bodies, transforming volition into reinforcement and action into conditioned response."[20]

It is precisely this kind of description of the power exercised by contemporary media and tech moguls that is at the heart of Sonia Livingstone's criticism when she articulates that "to theorize recent and profound changes, media scholars are reasserting monolithic accounts of power that tend to downplay or exclude audiences and the significance of the lifeworld."[21] When we settle for the accounts that reduce human subjectivity to an easily hackable and predictive model, we are simply taking for granted the behaviorist approach that media scholars like Merit de Jong and Robert Prey consider to be the *episteme* that grounds the development of platforms.[22] De Jong and Prey argue that platforms fueled by data and algorithms are based on a "behavioral code that promotes an impoverished view of what it means to be human."[23] We agree with these two scholars when they argue that "leaving this technical code unchallenged prevents us from exploring alternative, perhaps more inclusive and expansive, pathways for understanding individuals and their desires."[24]

The focus on platform power and its apocalyptic effects on society risks obscuring the investigation of what kind of agency, if any, still rests in the hands of the people, resulting in a disregard of individual agency in the discussion of the consequences of algorithmic culture and algorithmic infrastructures. Nancy Ettlinger[25] has pointed out that while key conceptualizations of algorithmic governance (including data colonialism and surveillance capitalism) are particularly strong in accounting for subjection and domination, they tend to overlook agency and leave resistance unattended, despite the proliferation of experiences that point to the contrary. In line with Ettlinger's reflection, this book casts resistance as part of an ecosystem of digital governance and recognizes the importance of situating forms of agency and resistance within the biased structures of domination and oppression that constitute the platform society. Our gaze is primarily oriented toward people's practices and encounters with algorithms, including the creativity and imagination mobilized and the challenges and obstacles faced every day by people while coping with algorithms. In this sense, we are following in the wake of studies on data agency,[26] data activism,[27] and everyday practices of living with data and algorithms.[28] These accounts, instead of focusing on top-down processes of datafication, dwell on the ways in which ordinary people and global social movements make sense of data from below, appropriating them for their own needs and purposes.

However, we are deeply aware that users' agency in the platform society is the result of both the constraints posed by the structures of platform power and their ability to exploit the affordances of platforms to their own advantage. In addition to studying the infrastructures and the effects of this growing power, we believe that it is necessary, as media scholars, citizens, and activists, to investigate the power still available to people to assert their autonomy of choice and find their own *dance rhythm* through the deafening chaos brought by the rise of the platform society.

We are convinced that media and data scholars need to work on both sides of the barricade. Hence, while our gaze is informed by the analysis of platform power,[29] it is at the same time oriented toward the forms of "audience/user appropriation" of this power. Michel Foucault famously argued that "where there is power, there is resistance, and yet, or rather consequently, this resistance is never in a position of exteriority in relation to power."[30] Power and resistance, for Foucault, always must be thought of together, as they are inseparable. Acts of resistance, rebellion, and sabotage are born of a response to existing systems of power, control, and domination. Platform power is thus inseparable from the ability of individuals to exert some sort of agency and resistance over it. This does not mean that platform power can be easily counterbalanced by these practices of resistance. Neither is it our intent to provide romantic and heroic accounts of forms of agency and acts of resistance: these acts, as we shall see, emerge laboriously amid the constraints posed by the platforms, may come from significantly different social and political formations, and can be enacted for pure profit and propaganda purposes. Yet we believe that zooming in on the strong link between power and agency, and between power and resistance, is necessary if we want to understand the complex grammar of contemporary algorithmic cultures.

ALGORITHMIC AGENCY

Our aim is to understand the practices that people enact to cope with algorithmic infrastructures. To make sense of them, we propose a comprehensive framework that will help us to realign attention toward the forms of agency that arise in various contexts as datafication penetrates systems of governance and political environments. Instead of a rejoinder of the old-fashioned debate about "media versus audience power," By proposing a framework in which algorithms and user agency are shaping each other in evolving and complex ways we follow Taina Bucher's consideration that, "while algorithms certainly do things to people, people also do things to algorithms."[31]

In so doing, we are not alone. Other scholars have already started to look at algorithmic media as a battleground of contesting actors. For instance, Julia Velkova and Anne Kaun questioned "the extent to which everyday media users are only subjects and victims of algorithmic power,"[32] while Jeremy Morris[33] observed that content producers, marketers, and users invent creative (and sometimes unauthorized) uses to take advantage of the platforms' affordances, both to achieve greater visibility and increase their profits. Furthermore, the geographer Rob Kitchin focused on the ways in which people "resist, subvert and transgress against the work of algorithms, and re-purpose and re-deploy them for purposes they were not originally intended."[34] These few examples show that algorithmic environments are much more contested than has been thought so far, and that the power exerted by them is never frictionless. Instagram "pods," Uber "surge clubs," attempts to fake personal workout loads and boost restaurant ratings, "Tinder scams," and "spoofing" of location-based videogames are just some of the dozens of algorithm gaming practices enacted every day by gig workers, fans, activists, and institutions of various kinds to make the algorithms work to their own advantage. We show that all those actions aimed at intentionally influencing algorithmic outputs represent manifold articulations of user agency in facing the power of the algorithms and the institutions that generate them—a kind of agency that we call *algorithmic agency*.

The sociologist Anthony Giddens defined agency as "the ability of human beings to make a difference in the world, that is, to exercise some sort of power."[35] To exercise agency, therefore, someone should be capable of doing things that affect the world in which they are embedded. The media scholar Nick Couldry developed a similar, but deeper, definition of agency as he underlined the centrality of reflexivity in the meaning of agency: "the longer processes of action based on reflection, making sense of the world so as to act within it."[36] Till Jansen,[37] while distinguishing human agency from the agency of algorithms, considered the former an evaluative and reflexive action and claimed that algorithms lack evaluative and reflexive agency. Combining these two latter definitions by Couldry and Jansen, we do not use "algorithmic agency" to refer to the agency of the algorithms, but rather to a user's "reflexive ability" to make the algorithms work to meet their own needs. We are aware that the algorithms will respond to any gaming attempt by recursively restructuring their results. Not only people do things to algorithms, but algorithms can affect people too: in fact, as the Italian sociologist Massimo Airoldi noted, algorithms are social agents that *agentically* make a difference.[38] The relationship between users and algorithms is thus a potentially infinite recursive one: users' agency is continuously restructured by algorithms, but they can also structure them back. The human-algorithm interaction is symbiotic. In fact, Gina Neff and Peter Nagy, while defining this relationship, proposed the concept of *symbiotic agency* to demonstrate "how agency is co-constituted in complex interactions between society and technology."[39] This *symbiotic* approach highlights the complex entanglement between human agency and artificial intelligence (AI). Giddens's structuration theory,[40] with its reflection on the duality of structure, highlights the process of mutual shaping between individuals and power structures. In a similar fashion, human agency and algorithmic infrastructures mutually shape each other. We thus conceive algorithmic infrastructures as Giddens conceived structure: both as "the medium and the outcome of the conduct it recursively organizes."[41] Therefore, algorithmic agency is *the reflexive ability of humans to exercise power over the "outcome" of an algorithm.* However, this agency is symbiotically embedded in the environment in which it is exercised; people exercise their agency while *acting upon* certain algorithmic outputs and, at the same time, by *reacting to* them. This symbiotic relationship happens within the boundaries of the affordances of algorithmic infrastructures. This means that human ability to exercise agency is shaped by the affordances of the platform and depends on the kind of power relations established by the platform. Yet, through our extensive ethnographic work, we will show how, even when they are stuck in asymmetric power relationships, people are still able to exercise some kind of agency.

Users' attempts to interfere with the outputs of algorithmic processes are often represented in a rather rigid, Manichean way that is either negative, like gaming attempts, or positive, like attempts to optimize user profiles. Yet, as we will see in chapter 2, these forms of agency can be shaped by different moral values and can be exerted by resorting to various kind of resources, either tactical or strategic. In some cases—but not always—these forms of agency can also be understood as more or less openly intentional forms of resistance to platform power.

ALGORITHMIC RESISTANCE

The forms of algorithmic agency that we illustrate cannot all be defined as acts of resistance to the power of platforms. In most cases, the pure exercise of one of the four ideal typical manifestations of algorithmic agency that we will describe in chapter 2 does not represent at all a form of resistance to this power. We argue that there is neither a clear distinction between agency and resistance nor a perfect superposition. Rather, we propose that the agency manifestations that we describe move along a *continuum* that goes from forms of agency openly resisting platform power to other forms of agency that have no intention of questioning or challenging such power. Let us clarify this point further by digging into the meaning of algorithmic resistance in relation to power.

The power that these platforms exercise is mostly invisible to ordinary users and is based on algorithms whose functioning is, to say the least, opaque. The emerging power accumulated by tech companies therefore closely resembles the idea of power described by Foucault: a diffuse, ubiquitous power that is accepted to the extent that it is hidden. Platform power is a "black-box" power. In fact, it is more easily accepted by those who are unaware of the functioning mechanisms of algorithms and their possible biases. Yet the fact that platform power is so pervasive and invisible to most people does not mean that people are hopelessly trapped inside the platform society. We cannot forget one of Foucault's most significant lessons, that where there is power, there is always resistance.[42] In this perspective, platform power is indissoluble from people's resistance. This does not mean that this power can be easily counterbalanced by the resistance of individuals to it, but that this power is exercised on bodies that are not passively subjected to it. If the power were "saturated" (e.g., totalitarian), we would be in the domain of domination. Yet domination can never be total. As George Simmel observed, "Even in the most oppressive and cruel cases of subordination, there is still a considerable measure of personal freedom."[43] Thus, the existence of everyday tactics of resistance offers "daily proof of the partiality of strategic control and in doing so they hold out the token hope that however bad things get, they are not necessarily so."[44]

Just as power is hardly given without resistance, acts of resistance in turn do not escape the dynamics of power. As David Courpasson and Steven Vallas rightly observed, resistance "is never as pure or pristine as a phenomenon as generations of Marxist theorists have hoped."[45] In resisting power, subordinate subjects also end up exercising power over their peers or other groups even more powerless than them. Foucault theorized the relationship between power and resistance, but most of his work has been marked by an effort to define power, leaving resistance in the background.[46]

To conceptualize resistance, we must turn to the work of the sociologists Jocelyn Hollander and Rachel Heinwhoner.[47] As they have shown, the definition of resistance is highly contested among scholars: it can take various forms, overt to covert, at different levels (macro and micro) and can be exercised with different degrees of awareness by the subjects. Some thinkers argue that to be defined as such, an act of resistance must be both intentional and recognized by the recipient, but this maximalist definition has been questioned several times. The American political scientist and anthropologist James Scott[48] had already shown that to be effective, the daily acts of resistance practiced by Malaysian farmers had to escape the eyes of the target (that is, the landowners).[49] Scott's farmers had good reason for not showing their defiant feelings to their employers. Other scholars contend instead that to be defined as resistance, an act must necessarily be intentional (i.e., understood as such by the subject who performs it), but we side with Courpasson and Vallas[50] in disagreeing with this view. There are acts of resistance that are not understood as such by the person who performs them but are instead recognized as such by the subject who receives them. The questions of the intentionality of the act of resistance are slippery: Who defines whether an act is conscious and intentional? And even if we could be able to establish the intentionality of an act, can we talk about resistance if this act has no effect on the target and nobody notices it?

Our understanding aligns with Courpasson and Vallas when they maintain that "an oppositional intention can by no means suffice as a valid indicator of resistance."[51] We thus conceive resistance as a dynamic phenomenon that can take many shapes and happen at various levels, even if the actor does not intend it as such. Different types of resistance depend on the ability of the subjects of the act of resistance, the targets of the act, or an external observer to recognize this act of resistance as such. As a result, there may be acts of resistance that are recognized as such only by those who make them, or exclusively by those who receive them (the target). Others are acknowledged only by an external observer or by all three actors mentioned here. The combination of these subjects (actors of resistance, target of the act of resistance, and an external observer) gives rise to the complex tapestry of resistance. Further, this multifaceted conception connects to another strand of research in resistance studies.

Taking up the work of Scott,[52] Mikael Baaz et al.[53] emphasize resistance as an act, regardless of the intent that moves it and the effects it can have, and show that, in addition to being an act that opposes power, it can be a productive act. According to them, an act of resistance can be such even if it is unconscious or if it does not achieve the expected results. Resistance is defined as "(i) an act, (ii) performed by someone upholding a subaltern position or someone acting on behalf of and/or in solidarity with someone in a subaltern position, and (iii) (most often) responding to power."[54]

We adopt this last definition of resistance and apply it to the realm of platforms and algorithms. Thus, when we talk of *algorithmic resistance*[55] throughout the following chapters, we intend to refer to (1) an act, (2) performed by someone upholding a subaltern position or someone acting on behalf of and/or in solidarity with someone in a subaltern position, and (3) (most often) responding to power *through algorithmic tactics and devices*.

RESISTING "TO" OR RESISTING "THROUGH" ALGORITHMS?

Reflecting on the mutual shaping between data and activism, the data scholars Davide Beraldo and Stefania Milan[56] have introduced a distinction between "data as stakes" and "data as repertoire." In the former (*data-oriented* activism), data are the "main stake in a hypothetical claim-making agenda,"[57] while in the latter (*data-enabled* activism), they are inserted in the repertoire of action of social movements and activists "alongside other more traditional forms of protest and civic engagement."[58] We apply this distinction to the realm of critical algorithm studies foregrounding algorithms as both *stakes* (resistance *to* algorithms) and *repertoire* (resistance *through* algorithms). In the former type, we find many political activists, organizations, artists, and critical scholars who openly resist the power of algorithms through collective actions, protests, art installations, and research that highlight the many risks that our society runs when it delegates its choices to AI systems (see also chapter 5). This is what we call resistance *to* algorithms. This resistance sheds light on negative effects of platform power (namely, the several biases that are often reproduced by the proprietary algorithms of digital platforms). In this case, algorithms represent the object against which artists, citizens, scholars, and activists' narratives and protests are oriented (algorithms as *stake*). For example, Mimi Onuoha and "Mother Cyborg" (also known as Diana Nucera) are two American artists who have written *A People's Guide to AI*, a text conceived partly as an artistic intervention and partly as a basic textbook for all citizens who want to learn more about the social consequences of AI-based technologies: "This booklet aims to fill the gaps in information about AI by creating accessible materials that inform communities and allow them to identify what their ideal futures with AI can look like."[59]

However, in addition to these forms of resistance *to* algorithms, there is a resistance to the power of platforms that is exercised *through* the algorithms themselves. Algorithms, in fact, can also be the tool *through which* citizens, workers, artists, critical scholars and activists exercise their protest actions (algorithms as *repertoire*). As Ettlinger clarifies, "Algorithms . . . can afford possibilities for resistance as much as for subjection."[60] Citizen practices that deploy algorithms as *repertoire*, including the creativity, resourcefulness, and difficulties that workers and activists face every day while coping with apparently "magic" decisions taken by an algorithm, can thus be interpreted as forms of algorithm-*enabled* resistance. As we will see in chapter 5, social movements' practices are now increasingly inserted and played out within the algorithmically defined environments of social media platforms that present both constraints and opportunities for activists who can also use algorithms for their own political purposes. Thus, algorithms can represent target objects of acts of resistance but they can also be tools through which people can challenge the power of platforms. This point is touched on by Ettlinger when she describes the forms of digital resistance that she calls "productive":

Productive digital resistance is algorithmic insofar as algorithms can be used as tools to enable digital subjects to develop new elements of the digital environment (e.g. apps, software, websites . . .) that target and subvert strategies-technologies-of repressive power, which may or may not be constructed with the aid of algorithms.[61]

Ettlinger labels as *productive* those forms of resistance that "make use of, or subvert, rather than reject or obfuscate, elements of the digital environment to serve digital subjects."[62] Among these forms of *productive* digital resistance, she includes the activities of hackers, civic "hacktivism," platform cooperatives, cloud protesting, and the use of art. However, this definition focuses only on those forms of resistance that come from politically conscious and tech-savvy citizens. On the contrary, we show that these forms of *productive* resistance are not only the domain of civil society, hackers, and contemporary artists, but they also can be found among gig workers and other people, who appropriate algorithms in an almost "situationist" way, as a "ready-made" object, as a weapon found on the street, on the way, and not as a tool built after a long design process. The forms of algorithmic resistance that we explore here are mainly examples of everyday microresistance, exercised by all those people who, despite not being completely aware of the power of platforms, incessantly invent ingenious ways of coping with it.

EVERYDAY FORMS OF ALGORITHMIC RESISTANCE

Although we should not romanticize these forms of microresistance, neither should we underestimate them. The aim of these actions is often not to subvert or question the power of platforms, as is the case with forms of resistance *to* algorithms. Often, there is no political awareness in these actions, no consciousness of being exploited or stripped of a highly precious and volatile value that quickly disappears into some tax haven. There is only an attempt to gain some small personal or collective advantage, to save time and money, to obtain temporary victories, to put a spoke in the wheels of the platforms. These actions of microresistance closely resemble the forms of "everyday resistance" described by James Scott in his study of Malaysian peasants opposing the so-called Green Revolution in agriculture. For everyday forms of peasant resistance, Scott intended to discuss "the prosaic but constant struggle between the peasantry and those that seek to extract labor, food, taxes, rents, and interest from them."[63] He had in mind the ordinary tactics of relatively powerless people: foot dragging, dissimulation, false compliance, pilfering, feigned ignorance, slander, arson, and sabotage. According to him, these actions required little or no coordination or planning; they often represented a form of individual self-help; and they typically avoided any direct symbolic confrontation with authority or with elite norms. But what does this everyday peasant resistance have in common with the tactics put in place by users all over the world to cheat or "game the algorithms" of online platforms? At first glance, the comparison seems misplaced: Scott's farmers sabotaged the combine harvesters, slowed the pace of the harvest, refused to take the place of other peasants, and hid part of the harvest in order to survive. The daily forms of resistance described by Scott were acts of survival, while the activities described in this book seem to be much more frivolous: a group of Korean teenagers who make their idol number one on Spotify's playlists, a group of Instagram microinfluencers exchanging "likes" and comments, and Tinder users disguising their profile to achieve more visibility.

Yet, in the next chapters, we will show that is possible to draw certain similarities between the tactics used by Malaysian farmers and platform users. These similarities can help us better understand the contemporary entanglement between platform power, user agency, and possible forms of resistance to this power. We are also aware of the huge historical, cultural, and political differences between the two contexts. Like Scott, we will be careful not to romanticize these practices. Even though he himself repeated several times that "it would be a grave mistake to overly romanticize the weapons of the weak,"[64] he was still criticized for saying this. As we will see, these practices can be put at the service of different intentions, some of which are not necessarily positive or morally acceptable to the majority.

This book contributes to foregrounding algorithmic agency and resistance not as episodic, but as ingrained into the very fabric of our everyday experiences. They are *ordinary* acts. As the British media scholar Roger Silverstone noted, it is in the realm of the mundane and the ordinary that individuals engage with hegemonic power structures. The "power of the ordinary" is grounded, according to Silverstone, "in the capacity of subjects . . . to appropriate and make their own meanings out of the stuff of an imperfectly hegemonic system, and in such appropriation and with varying degrees of consciousness, to oppose it."[65] In this sense, the practices of algorithmic resistance that we documented are *banal* (i.e., much more common and obvious than we might think). We insist on the *banality* of these acts because they are harnessed by various types of actors for disparate purposes, which can turn themselves into weapons available to both progressive agendas and extremely reactionary ones. They constitute the daily essence and grain of algorithmic life as much as platform power is. As our lives are progressively shaped by algorithms in all spheres and domains of social life, platform power, algorithmic agency, and resistance complement each other.

In chapter 2, we go beyond explanations based on the simplistic "gaming" versus "optimization" opposition and instead rely on the concept of the moral economy to frame these acts as being invested with different moral values. Bringing together the theory of the moral economy and the tactical/strategic dimensions proposed by Michel de Certeau[66] will help us to sketch a complex, multidimensional understanding of how multiple actors resist algorithms in the platform society.

2

THE MORAL ECONOMY OF ALGORITHMIC AGENCY

INTRODUCTION

While doing research for this book, we came across several texts that contained moral judgments from both platforms and users. We found that the discursive regimes used by platforms to regulate the behavior of their users and those deployed by users among themselves were often steeped in different moral judgments about both user and platform behavior. Platforms exercise their agency on users through the mediation of algorithmic infrastructures, such as recommendation algorithms. Each platform also implements its own terms of service (ToS) that govern users' behavior, stipulating what they are allowed and not allowed to do. However, users are not programmed robots, so they do not always obey or comply with these rules. As we will see in the following pages, to enforce the ethical guidelines that are encoded in the ToS, platforms resort to precise discursive regimes that paint the behaviors that violate their ToS as negative, artificial, or even immoral. At the same time, however, users frequently develop their own alternative ethical codes.

This book captures the stories of resistance to and subversion of algorithmic infrastructures that people develop across a variety of situations in response to the affordances, regulations, and discursive regimes of digital platforms. Before diving into these stories, in this chapter, we lay out the theoretical foundations of the book and describe the conceptual framework that has emerged from our field research, our extensive review of the literature, and our analysis of a large corpus of articles. This discussion allows the reader to understand where we situate our research in the context of contemporary studies on platform power, agency, and resistance and, more importantly, provides them with interpretative keys to read and decode the rest of the book.

More specifically, this chapter introduces two conceptual pillars of our understanding of platform power and algorithmic agency. The first is represented by the adoption of the concept of the *moral economy*,[1] which will be used to demonstrate that users' agency is shaped by moral values that might compete and even collide with those of the platforms. The moral economy enables us to make sense of a vast and apparently different array of practices that are usually described as activities aimed at "gaming the algorithms," but can be better intended as forms of reflexive agency enacted by citizens and social groups to "live with" algorithms.

The second pillar consists of the introduction of a further dimension in the four manifestations of algorithmic agency that we have identified. In addition to having a moral dimension, we show how algorithmic agency can have either a *strategic* or *tactical* dimension, depending on the resources available to the actor that exercises it. These pillars define our theoretical framework and allow us to disentangle the multidimensional nature of algorithmic agency by (1) articulating this agency along a continuum shaped by two competing *moral economies*: the *user* moral economy and the *platform* moral economy; and (2) understanding this particular form of agency as either *tactical* or *strategic*. Based on our conceptual framework, we identify and flesh out four manifestations of algorithmic agency at the end of this chapter—namely, *strategic algorithmic agency aligned or not aligned with the platform moral economy* and *tactical algorithmic agency aligned or not aligned with the platform moral economy*—which then will be explored in detail throughout the book (see figure 2.1).

2.1   Competing moral economies and tactical/strategic dimensions of algorithmic agency.

THE MAKING OF THE MORAL ECONOMY

This was the second year of the scarcity. In the preceding one, the provisions, remaining from past years, had supplied in some measure the deficiency, and we find the population neither altogether satisfied, nor yet starved; but certainly unprovided for in the year 1628, the period of our story.[2]

This is the beginning of chapter 12 of *I Promessi Sposi* (The Betrothed), one of the most popular novels ever written in Italian, first published in 1827. After Dante's *Divine Comedy*, Alessandro Manzoni's *I Promessi Sposi* is probably the most loved (and hated) novel by Italian students, who were compelled to read it at school (whether they wanted to or not). In this chapter, Manzoni brings to life, in the style of a historical novel, a real event: the bread revolt that took place in Milan in November 1628 under Spanish rule. At that time, after a long famine, bread had become a precious good.

The Spanish high chancellor Antonio Ferrer had imposed a political price on bread, which was challenged by the bakers as being too low. The chancellor, concerned about the bakers' complaints, appointed a council, and the price of bread rose again. While this measure benefited the bakers, it aroused the discontent of the population, which immediately began to mobilize. Manzoni, then, dwells on the sacking of the Forno delle Grucce, the most popular bakery in the city at the time.

He placed the protagonist of his novel, Renzo Tramaglino, amid the crowd assaulting the Milanese bakery and painted him being carried away by the angry mob. Manzoni, who as an adult had become a fervent Catholic with liberal ideas, saw in this wild crowd the manifestation of the irrational instincts of the population. Manzoni describes the transformation of poor, starving "Christians" into raging beasts ("ferocious and bloodthirsty").[3] The crowd assaulting the bakeries is portrayed as a herd hunting for prey, governed by emotion and transformed into a mass of burglars.

This way of describing a hungry crowd of people as a ferocious and irrational herd is exactly the object of the famous British historian Edward P. Thompson's critique in his 1971 essay, "The Moral Economy of the English Crowd in the Eighteenth Century." In the opening paragraphs of this essay, he criticizes the "spasmodic view of popular history," according to which, at least before the French Revolution, "the common people can scarcely be taken as historical agents."[4] Thompson aimed at revealing "the historical agency of 'the crowd' against 'spasmodic views of popular history' that naturalize and reduce people's actions to automatic quasi-biological responses to hunger."[5] In this spasmodic view, which coincides exactly with Manzoni's perspective on the Milanese bread riot, the crowd's actions are understood as nonpolitical because of their spontaneous, almost instinctive nature.

Thompson's view, however, is extremely critical of this understanding of the English crowd. According to him, the practices of attributing value to basic goods that were emerging alongside the nascent market economy during the eighteenth century repeatedly clashed with those associated with a preexisting moral economy, sparking food riots and other forms of public protests. He proposed a new interpretation of the logic of the riots. He understood them as expressions of the crowd's moral vision of the economy, which in their eyes legitimates their uprisings. This legitimacy was grounded "upon a consistent traditional view of social norms and obligations, of the proper economic functions of several parties within the community, which taken together, can be said to constitute the moral economy of the poor. An outrage to these moral assumptions, quite as much as actual deprivation, was the usual occasion for direct action."[6]

The men and women in the crowd believed that they were defending traditional rights or customs. Thompson intended the food riots as a "highly complex form of direct popular action, disciplined and with clear objectives":[7] the poor acted not only to seek sustenance—forcing merchants, millers, or wealthier farmers to sell grain (or bread) at what they considered the "customary" or "moral" price—but also to punish those they considered "profiteers," for acting according to market logic was seen as predatory.[8]

The British historians Andrew Charlesworth and Adrian Randall remind us that "frequently, flour and grain were destroyed in a public demonstration of communal punishment of those deemed guilty of immoral practices."[9] Yet Thompson didn't claim that food rioters were more moral than Adam Smith's followers: in fact, he later explained that he "was discriminating between two different sets of assumptions, two different (moral) discourses."[10] This distinction allows us to argue that each model of economy brings with it a different set of moral values, rights, and customs, and thus there is no single moral economy. In fact, the sociologist Andrew Sayer, in a 1999 essay that revisits Thompson's work, underlined that the concept of the moral economy refers to the way in which "all economies are suffused with values and beliefs about what constitutes proper activity, regarding rights and responsibilities of individuals and institutions, and qualities of goods, service and environment."[11] All economies, then, are moral.

Since Thompson first mentioned this concept, many other authors have appropriated it. In particular, the popularity of the concept of the moral economy owes much to the anthropologist and political scientist James Scott and his 1976 book, *The Moral Economy of the Peasant*, in which he describes the set of moral values, habits, and beliefs that shaped the Vietnamese and Burmese peasants' view of the economy. According to Marc Edelman, these values included the peasants' notions of "just prices" (including "just" rents and taxes), "as well as other sorts of entitlements, such as access to land, gleaning and fishing rights, and forms of reciprocity that linked peasants with elites and with each other."[12]

With his work, Thompson showed us that "the market remained a social as well as an economic nexus":[13] the English crowds of the eighteenth century were endowed with their own agency, and this responded to a precise moral economy, as opposed to the market economy.

Thompson thus argues that human agency can be influenced by different moral economies. Distinguishing these economies is important because it helps us understand the different reasons and values that drive human agency. Building on Thompson, in the next section, we will distinguish the moral economies that drive algorithmic agency.

THE COMPETING MORAL ECONOMIES OF ALGORITHMIC AGENCY

Our first proposal, therefore, is to distinguish the various forms of human agency available in coping with algorithms—what we called "algorithmic agency" in chapter 1—along a continuum shaped by moral values that sees, at its two extremes, two competing moral economies: *platform* and *user* moral economies.

PLATFORM MORAL ECONOMIES

Platforms are not neutral artifacts. As the American social scientist Langdon Winner noted, every technical artifact has political qualities.[14] The affordances of each artifact are designed to favor some actions at the expense of others. Artifacts also have gender, in the sense intended by the science and technology (STS) scholars Anne Jorunn Berg and Merete Lie, because "they are designed and used in gendered contexts."[15] Whether or not they are aware of it, designers transfer their values to the technologies that they design. For example, the affordances of Facebook's content moderation ecosystem discourage users from posting an image of a female human's nipple: if they do, the image is immediately banned.[16] Platforms express the moral values of those who created them. The field of research into the moral values expressed by platforms represents a promising new area of inquiry: the British political geographer Louise Amoore, for example, understands algorithms as ethical-political entities that generate "their own ideas of goodness, transgression and what society ought to be."[17] The American sociologists Jenna Burrell and Marion Fourcade argued that "algorithms are transforming the very nature of our moral intuitions—that is, the very nature of our relations to self and others—and what it means to exist in the social world."[18] The Jewish new media scholar Limor Shifman and her colleagues have started a research project around "digital values."[19] They argue that "technological systems are bound up in our social, ethical, and moral worlds,"[20] and "corporate value statements, policy documents, financial disclosures, and public statements contribute to the corporate construction of values."[21] By analyzing the affordances and the terms of service of a platform, the public speeches of its founders, and their internal documents, we can infer the set of values that any platform embodies. In fact, as Bruno Latour had already observed, "technology is society made durable"[22]—that is, the values of a society are crystallized in a specific form of technology.

But what kind of values do digital platforms embody, exactly? Some might argue that platforms have no morals, that they only follow profit. Yet we should avoid such a simplistic view. Facebook's censorship of the nipple image shows us that things are more complicated than that. Platforms do have morals (that is, they have a clear vision of how users should behave in their digital environments). In fact, those who do not behave themselves are expelled from the platform—"deplatformed"—as happened to President Donald Trump, or "shadow-banned."[23] When we talk about the moral economy of platforms, then, we mean a precise set of values that shapes the kinds of actions that users are allowed to take. These moral economies are not the same for all platforms, just as terms of service are not the same for everybody. As José van Dijck and her colleagues noted, competing ideologies of capitalism and democracy coexist in the platform society.[24] Public service–oriented European platforms are aimed at protecting user privacy and developing a data policy that conceptualizes data as the digital commons.[25] US commercial platforms, European public service platforms, platform cooperatives, and Chinese state-controlled platforms are shaped by different sets of values and produce different moral economies. Not only are there differences between the moral economies of platforms emerging from diverse ideological contexts, but there also are differences within the companies themselves that produce the platforms: not all actors—interaction designers, software developers, business managers, or others—think alike about the kinds of things that a platform should enable. As Blake Hallinan and her colleagues point out, "The construction of platform values is not a smooth, unified process; platform values are contested, with different ideas of the desirable playing out among actors through various modalities."[26]

There is certainly still much to be done in the study of platform values and platform *agency*. To simplify our argument, however, in this book we will mainly refer to the moral economy of the commercial platforms employed by the users whom we observed in our fieldwork. The moral economy of these platforms, from Deliveroo to Instagram, from Twitter to Uber to Airbnb, is mainly centered on neoliberal values such as the free market, individualism, consumerism, data extractivism, optimization of performance, self-entrepreneurship, meritocratic ideology, and competition among users. According to this view, tech companies consider it just, fair, and morally acceptable to extract personal data for profit or to encourage and reward competition and self-entrepreneurship. These values constitute the heart of the moral economy of platform capitalism, the currently hegemonic form of communicative capitalism.[27] For the sake of simplicity, we will call it the "platform moral economy" because it is embodied by most commercial platforms.

USER MORAL ECONOMIES

The moral economies through which platforms exercise their power are normally taken for granted by users. Platform moral economies are, in most cases, perceived as "natural" by users, and not as the result of a specific corporate "culture." When someone does not comply with the rules of behavior set by the platform, the first to intervene are often the users themselves, even before algorithms and human content moderators can notice the violation. It is very likely, in fact, that users who are sanctioned by Twitter or Facebook for a tweet or post deemed offensive were first reported by other peers. Most platforms, in fact, encourage users to flag problematic content and behavior that are subsequently evaluated by human content moderators.[28]

In most cases, when everything goes smoothly, users do not question the algorithms that lie behind the stream of posts, images, or jobs recommended by the platforms. On the contrary, many of them, although they are now aware of the existence of algorithms, are thrilled to be able to save time and be relieved of the "burden of choice."[29] Even among platform workers, many people are satisfied with the work they do. Among the online food delivery couriers whom we interviewed, for example, there were some who said they were happy to have left the factory, even though they had a stable employment contract because they felt freer on their bikes and more able to control their time. Yet, as a survey of nearly five thousand gig workers in fifteen countries showed, the experience of platform work worldwide is shaped by mixed emotions and dim prospects. Satisfaction levels varied widely from country to country, but they were generally worse in lower-income countries, where workers tended to feel more worried, unsafe, tired, or angry.[30]

Among those who feel less satisfied with the benefits of platforms, discontent is beginning to grow, and negative feelings and criticisms of the platforms' business models are emerging. While it is true that most users naturally accept the values embodied in the platforms that they use, there are also users who begin to question these values, or even challenge them more openly. These users do not refuse to use platforms (or they cannot afford to refuse them, as in the case of some platform workers), but at the same time, neither do they love them, and, indeed, they develop strongly critical views of them. The set of sentiments and values that coalesce around these critical views can be considered an alternative moral economy to that of platforms.

Within this moral economy, users have a different view of what is legitimate and what is not, and this view may also collide with that inscribed in the ToS of the platforms. This "oppositional" kind of moral economy can take various shapes: it can be centered on "social/collectivist" values (mutual support, cooperation between users, political awareness, struggle organization), or it can lean toward "entrepreneurial solidarity," as the Filipino scholars Cheryll Ruth Soriano and Jason Vincent Cabañes noted,[31] in which users support each other and share knowledge and "tricks," but their aim remains that of optimizing their online behavior and does not go as far as organizing politically to challenge the platforms. In other cases, it can even be based on a "free rider" or "pirate" ethic, which seeks values of individual profit, does not recognize the meritocratic ideology, and has no problem with breaking the rules of the platforms (this is the case with many platform scams).

As in the case of the moral economies of platforms, there is no single-user moral economy. On the contrary, within these two families of moral economies, the value sets are multiple and the differences among them highly nuanced. For the sake of simplicity, however, we will refer to this second group of moral economies with the ideal-typical concept of the "user moral economy." It brings with it various values and rules of behavior that users recognize as moral, legitimate, and worthwhile, even if not in line with those expressed by the platforms.

Why mobilize the concept of the moral economy to make sense of algorithmic agency? Because it allows us to defuse the paternalistic rhetoric inscribed in the gaming versus optimization distinction, showing how the assignment of a morally negative (gaming) or positive (optimization) value to a certain practice depends on the type of moral economy embodied by who assigns it.

GAMING IS IN THE EYE OF THE BEHOLDER

All those practices that resist, subvert, and transgress the computational work of algorithms are generally described by the media and by the platforms themselves as forms of "gaming the system"[32]—that is, practices that intentionally seek to manipulate algorithmic computations to their own advantage using means that are not legitimate, or at least not recognized as such by the codified rules of the platforms (i.e., ToS, community standards, or creator and user guidelines).

To put algorithms at work to reach their goals, users not only perform actions that platforms consider inauthentic, such as putting a Fitbit device on their dog to artificially boost their daily running performance, but also try to optimize their behavior according to the rules of the algorithms themselves. They do this to obtain greater visibility on social media, for example. Optimization practices, then, represent the clean face of users' desire to make the algorithms work to further their own interests.

Optimization and gaming are remarkably similar practices since both are intentional efforts to interfere with the results of technical systems. What distinguishes them is the means used to achieve this objective. In the case of optimization, these are considered legitimate practices (i.e., allowed by the ToS of the platforms and even supported by the platforms themselves). In the case of gaming, they are seen by online platforms as illegal practices that openly violate ToS or other codified platform rules.

Yet what seems to be a clear distinction represents instead an extremely slippery ground. The artificiality of this distinction, in fact, has been highlighted by many media and STS scholars.[33] The platform scholar Thomas Poell and his colleagues defined algorithmic gaming as "third-party algorithmic optimization tactics labeled as illegitimate by platforms."[34] The boundary between optimization and gaming is not so easy to draw because the inscription of a practice within the boundaries of the gaming or the optimization field is not a neutral process, but one that varies greatly depending on the perspective from which the phenomenon is observed. Poell and colleagues argue that the boundary between what is considered by platforms to be either legitimate or illegitimate constantly shifts, and what is initially understood as "optimization" can quickly transform into "gaming," and vice versa.[35]

We could say that gaming *is in the eyes of the beholder* (that is, it depends on the subject who is defining its boundaries). If we assume the point of view (and the moral economy) of the platform, some practices will be labeled as "gaming efforts." If we assume the point of view of the users, the same practices could be interpreted as legitimate attempts to interact with the platforms to gain more visibility.

The American media scholar Caitlin Petre and her colleagues brilliantly argued that platforms establish, maintain, and legitimize their institutional power rightly through a continuous redefinition of the boundaries of what is legitimate or not: "What is deemed 'acceptable' versus 'unacceptable' user activity is situated within ever-evolving cultural practices and power relations."[36] They also recognized that the "boundary between what platforms deem legitimate strategic action and illegitimate algorithmic manipulation is nebulous and continually shifting in accordance with platforms' business strategies."[37] This moral boundary-drawing fosters a dynamic that Petre et al. call "platform paternalism," an orientation that "not only imbues platforms with structural and economic power, but moral authority as well."[38]

For example, one artist who was intentionally trying to manipulate to his own advantage Spotify's royalty payout scheme by playing his own music for long periods of time received the following notice from the digital distribution service Distrokid: "[Your song was removed] because the song was streamed a massive number of times, but by a tiny number of people."[39]

Hence, a certain practice constitutes gaming only if the platforms themselves define it as such. They exercise their "paternalistic authority" to police the audience/users, render them compliant with their ToS, and channel them toward the "appropriate" user behavior (or, to use an old-fashioned category in media studies, "dominant hegemonic decoding"), as expressly designed to increase their profits.

Instead of focusing on the supposed legitimacy or inauthenticity of the gaming and optimization practices, the American STS scholar Malte Ziewitz proposed that concerns about the prevalence of gaming might be better understood as forms of moral regulation. "What counts as 'gaming,'" Ziewitz wrote, "is not given in advance, but needs to be established, navigated and negotiated in specific situations."[40] Building on Ziewitz's idea of the need to rethink the boundaries between optimization and gaming practices as shaped by moral concerns, we propose to understand these practices as driven and shaped by different "moral economies."

Just as the British historian Edward P. Thompson argued that the source of the social unrest that affected the UK between the eighteenth and nineteenth centuries lies in the tension between two models of economies—the moral economy of colliers, artisans and the poor and the market economy—we propose that the key to understand the gaming versus optimization distinction lies in the tension between the moral economies of the users and those of the platforms. The division between gaming and optimization is only good at revealing the power dynamics behind these labels. From our point of view, any action taken by users on platforms, rather than being considered a gaming or optimization strategy, should be considered an action shaped by particular moral economies. What is intended as gaming by someone can be interpreted as optimization by someone else. However, platforms have much more power than users to determine what is gaming and what is not.

MOBILIZING THE CONCEPT OF MORAL ECONOMY

So far, the concept of moral economy has not met with much success in media studies and is almost absent from more recent platform and critical algorithm studies. The term "moral economy" has been employed only sporadically.[41] The British media scholar David Hesmondhalgh recently suggested that "a moral economy approach might reinvigorate approaches to the media and culture,"[42] while Henry Jenkins and his colleagues[43] compared the moral economy of users who share and download "pirate" content from *peer-to-peer (P2P)* networks, considering it a legitimate practice, with the opposite moral economy of film production companies, who publicly framed film piracy as an immoral and criminal practice. The concept of moral economy has been used, with a different meaning, by the British culture and media scholar Roger Silverstone and his colleagues[44] to describe the moral dimensions of the acts of appropriation and domestication of communication technologies in the household. But apart from these well-known examples mentioned so far, the frame of the moral economy has been sparsely mobilized. We therefore propose to interpret the practices that aim at interfering with the work of algorithms as being informed by competing moral values, even when these practices are aimed at obtaining an economic advantage, such as click frauds, or a political advantage, such as bots and computational propaganda that intoxicate public discourse by producing artificial numbers of opinions in favor of a specific political agenda.

The English crowds of the eighteenth century rebelled against the rules of the free market that set the prices of basic needs. Riots exploded when people realized that bread was scarce, and therefore expensive. It was the rise in bread prices, induced by the free market, that generated the riots, which according to Thompson's interpretation, were lucid attempts to reestablish a fair price for essential goods. For British food rioters, bread could not exceed certain price thresholds because that was not fair. Similarly, for Instagram users, posts cannot fall below a certain number of "likes," or for music fans, their favored band cannot be excluded from a Spotify playlist, or for an Uber driver, the price of a ride cannot fall below a certain threshold, regardless of the organic conditions of the demand-supply relationship. From the platform users' point of view, interventions aimed at interfering with the algorithm to snatch a more equitable price are considered legitimate. The contested price, in our case, can indicate either a monetary value, as in the case of an Uber ride, or an apparently more intangible value, such as online visibility in the case of a content creator on Instagram.[45]

Any action aimed at restoring a fairer distribution of these goods and values is thus considered morally acceptable by the users because their moral economy does not completely align with the ideology of the free market coded into the algorithms of Silicon Valley. In other words, users do not always passively accept the implicit ideological discourse encoded into the platforms that they use every day. Sometimes they contest, negotiate, or even subvert that discourse.

Some authors, such as the media scholars Stine Lomborg and Patrick H. Kapsch,[46] have demonstrated the existence of different forms of "decoding" the power and meaning of algorithms and compared these forms to the three ideal types of decoding (dominant/hegemonic, negotiated, and oppositional) proposed by Stuart Hall in his classic 1973 study of the audience reception of television discourse.[47] Other media scholars, such as Ignacio Siles and his colleagues,[48] have also shown that users—in their case, musicians on Spotify—develop logics deeply similar to Hall's decoding categories to make sense of their relationship with the platforms. These authors argue that platform users react in clearly different ways to algorithmic power: some adhere to its hegemonic code, others negotiate it, and still others even resist it.

Our idea of the existence of competing moral economies goes in the same direction as these studies: we could also say that users aligned with the moral economy of the platform undergo the dominant discourse of it, while users who oscillate between complete adherence to the moral economy of the platforms and partial rejection of it respond to a form of negotiated decoding. And finally, those who do not adhere to the moral economy of the platforms and develop their own are like Stuart Hall's viewers, who develop an oppositional decoding of television discourse. Hall's lesson is that every power structure is a battlefield. Popular culture is a battlefield; media discourse is a battlefield. What we are arguing here is that algorithmic power is also a battlefield. Within this field, individuals perform actions that respond to different moral economies, which may even be partially misaligned or completely opposed to those embodied by platforms. And when users are not aligned with the moral economy of the platform, they can develop practices like those deployed by the English crowd studied by Thompson: "gaming" an algorithm can be a profoundly rational form of (digital) rioting, too.

If the English peasants of the eighteenth century rioted in the name of an unfair price of bread according to their moral vision of the economy, the users of the platforms riot, resist, or just counteract in the name of an unfair distribution of online visibility or better job conditions in the gig economy.[49] They fight against precarization and casualization, and for decent living wages, according to their moral vision of the platform economy. As Ziewitz reminded us, "For those subject to the system, trying to optimize their own appearance is often the only way to reclaim a degree of agency in a potentially oppressive setting."[50]

Thus, the manifestations of algorithmic agency that we will describe in this book can be shaped by different and competing moral economies.

STRATEGIC VERSUS TACTICAL ALGORITHMIC AGENCY

This moral dimension is key to understand algorithmic agency, but it is not sufficient because not everybody is able to exert agency over algorithms in the same way. What platforms and media call "gaming" efforts require information, time, monetary and technical resources, and, often, orchestration of collective actions. Hence, some will be in a better position to game the algorithms than others.

It makes a difference whether the attempt to game an algorithm is undertaken from "above," so to speak—by a state, an institution, a corporation—or conversely, from "below" (by a group of adolescents, a single individual, or a social movement). It is not enough to claim that all the manifestations of algorithmic agency that are labeled as "gaming" or "optimization" by platforms are better understood as being informed by competing moral values. This frame does not consider potential inequalities in the distribution of power (that is, access to digital devices, economic resources, technical knowledge, expertise, or digital literacy) to resist algorithms or to subvert their outcomes.

Our second proposal, therefore, is to add a further dimension to the moral dimension of algorithmic agency, which brings into the equation the type of power held by subjects that enact algorithmic agency. This further dimension is articulated along a second continuum between two poles: *tactical algorithmic agency* and *strategic algorithmic agency*.

Our use of the concepts of tactical and strategic action is informed by the French scholar Michel de Certeau.[51] The setting of strategy, noted de Certeau, is always the purview of power. Strategy presumes control and a subject with "will and power."[52] In contrast to strategy, de Certeau characterized tactics as the purview of the nonpowerful. He understood tactics not as a subset of strategy, but rather as an adaptation to the environment that has been created by the strategies of the powerful. To explain this difference, de Certeau made the example of urban planning strategies and the tactics employed by inhabitants to circumvent them. Walking in the city is used by the French scholar as an iconic example of a tactic because the designers of urban spaces cannot entirely predict the ways that people will move around them. For example, when streets are designed without considering the needs of pedestrians, they will develop shortcuts and alternative routes through terrain that was not planned to be crossed on foot. Further, de Certeau wrote that "the space of the tactic is the space of the other,"[53] meaning that tactics always operate in a place defined by the strategies of the powerful: the pedestrian walks in the space designed by urban planners, in a *foreign* terrain. In this territory, the pedestrian acts like a poacher, developing tactics to adapt or subvert the affordances strategically designed by the urban planner to meet their needs.

This distinction between tactics and strategies remarkably contributed to our understanding of everyday life, but many scholars have pointed out that these concepts have remained vague and ill defined in de Certeau's work and are open to many possible interpretations. The cultural theorist Ian Buchanan, in his understanding of de Certeau's intellectual legacy, contributed to better defining these two concepts. According to him,

The essential difference between the two is the way they relate to the variables that everyday life inevitably throws at us all. Strategy works to limit the sheer number of variables affecting us by creating some kind of protected zone, a place in which the environment can be rendered predictable if not properly tame. Tactics, by contrast, is the approach one takes to everyday life when one is unable to take measures against its variables.[54]

Tactics are materially ephemeral and fragile and are as much in danger of being swept away or submerged by the flow of events as they are of breaking through the dams that strategy erects around itself. Why is this distinction so central for us? We argue that every form of agency, as well as being inspired by different moral economies, can be exercised according to a strategic plan or in the form of everyday tactics. If we apply this distinction to algorithmic agency, we can also distinguish between *tactical* and *strategic* algorithmic agency.

Now, what happens if we cross de Certeau's distinction between tactical and strategic dimensions with Thompson's theory of the moral economy? We obtain the theoretical framework that lies at the heart of this book. By articulating these two dimensions (i.e., the different moralities of algorithmic agency and its tactical vs. strategic manifestations along two axes), we have drawn a matrix that is able to foreground all the possible nuances of algorithmic agency. Our conceptual framework is constituted by four quadrants emerged from the intersection between the moral economy axis and the strategic/tactical axis (see figure 2.1 earlier in this chapter)

We are not the first scholars to frame as "tactical" the types of alternative uses of the algorithms that users perform. Media scholars such as Julia Velkova and Anne Kaun, for example, "foreground the significance of mundane user encounters with algorithms through which users can develop tactics (see de Certeau, 1984) of resistance through alternative uses."[55] Yet they focus only on the forms of explicit algorithmic resistance that they call "media repair practices." Recently, de Certeau's thought has become fashionable again among all those scholars who are not satisfied with the dystopian narrative of a monolithic platform power: Justine Gangneux[56] talks of *tactical agency* to describe the way in which young people engage and disengage with WhatsApp and Facebook Messenger. Tanya Kant also drew on de Certeau to understand how web users cope with algorithmic infrastructures and explicitly described such users as "algorithmic tacticians," referring to all users engaged in "maneuvering within, against and through algorithmic anticipation."[57] At the very end of her book, Kant briefly discusses algorithmic tactics, in line with what we call *tactical algorithmic agency*. However, this type of agency is only one of two ways in which algorithmic agency can manifest itself. There is also a second type of agency that users can exercise on algorithms—one that we call *strategic algorithmic agency*.

We consider as strategic all those manifestations of algorithmic agency enacted by institutions, governments, national states, corporations, think

tanks, lobbies, public relations companies, or even individuals or users that
have social, economic, and cultural capital at their disposal to interfere with
the work of the algorithms according to long-term strategic visions. To act
strategically means having a high availability of time, money, and expertise
and being able to rely on long-term plans. To act *tactically*, on the contrary,
means having a low availability of time, money, and expertise and being
able to rely only on short-term plans. Strategic practices too, such as Chinese or Russian computational propaganda,[58] represent nonpassive forms
of cohabitation with the output of platform algorithms, even if their aim,
in this case, is oriented to the strengthening of one's own political or economic hegemony.

All the manifestations of *strategic* algorithmic agency that we have
mapped during our research show recurrent patterns in the fields of politics, gig economy, and cultural consumption: they are activities that rely
on considerable amounts of money, are more usually fueled by a team of
people who work full time, and are often supported by the deployment
of bots. The agency of those involved in these activities is augmented
by computational power and propelled by economic capital and a deep
knowledge of the functioning of algorithmic infrastructures.

Most people, however, cannot afford these resources, making do with
devising small-scale, short-term tactics. On the other side, then, we
consider as *tactical* all those manifestations of algorithmic agency that
come from subaltern agents, such as ethnic, linguistic, and gender (intersectional) minorities; from social groups or individuals excluded from
democratic participation; or from users, fandoms, cultural producers, and
social movements that do not possess the necessary economic capital to
invest money in the promotion and marketization of their content. Even
if they lack computational and economic resources—or perhaps for this
very reason—the *tactical* actions individually or collectively organized by
platform users are incredibly varied and can reach high levels of complexity and sophistication.

The main difference between tactical and strategic algorithmic agency
lies in the quantity and quality of resources available (time, money, and
expertise) to users to exercise their agency. To "do things to algorithms,"[59]
a user or a group needs to have access to various types of resources: a rough

(at least) "algorithmic imaginary"[60] about the platform that they want to influence, economic capital that can be spent to promote their content or pay someone to orchestrate a collective action to have an effect on the platform's algorithms, a strategic plan that includes a schedule of several actions occurring over a period of time, and a network of peers that they can mobilize quickly. Different endowments of all these resources afford users to exercise their agency with different degrees of intensity. Various social actors have diverse amount of "time, expertise, and capital"[61] to spend on expensive optimization or gaming strategies.

When users have neither economic resources nor a long-term plan and can rely only on their limited network of social relations, the sphere of action of their agency will be highly limited. However, we should not consider these two forms of algorithmic agency as dichotomic and immutable over time. The urban planner Lauren Andres and her colleagues[62] have rightly noted how in de Certeau's *The Practice of Everyday Life*, "the boundaries between strategies and tactics seem immutable,"[63] while in other writings by him, there seems to be room for a more nuanced distinction. According to Andres and colleagues, this ambiguity is beneficial because "it opens the potential for tactics and strategies to be seen as a continuum rather than opposites."[64] In the framework depicted in figure 2.1, we have adopted this interpretation: the boundaries between the forms of strategic and algorithmic agency are neither clear cut nor stable in time. In everyday life, we can come across a broad spectrum of actions with varying degrees of strategic or tactical resources. User practices move along a continuum and their position is not fixed over time: as soon as a social movement, a group of users of workers in the gig economy, or even a single individual is able to accumulate more cultural, economic, and social capital, their agency will be able to generate more strategic, effective, and long-lasting actions.

FOUR MANIFESTATIONS OF ALGORITHMIC AGENCY

These two forms of algorithmic agency can unfold within the moral economy of the platform or within that of the user, or at the intersection between the two, as shown in figure 2.1. That is, users can either exercise their agency (both tactical and strategic) by accepting the moral economy of the platform without questioning it, or they can partially or totally reject it and act in ways that violate the ToS of the platform. These practices are normally considered by the platform to be gaming attempts. Yet, from the users' point of view, they could be instead considered fair and just because the users are acting according to their own personal moral economy, which may be partially or totally alternative to that of the platform. By introducing the concept of moral economy within the debate on the agency available to users in coping with algorithmic power, we can better understand the nuanced and multidimensional affordances of this agency. Finally, as sketched in figure 2.1, we can thus envisage four possible manifestations of algorithmic agency. This framework foresees four ideal types of algorithmic agency, but in everyday life, the differences between these ideal types are much more nuanced and the boundaries between competing moral economies much more blurred. In the next sections, we introduce four brief sketches of these four manifestations of algorithmic agency.

STRATEGIC ALGORITHMIC AGENCY ALIGNED WITH
THE PLATFORM MORAL ECONOMY

Seohyun is twenty-one years old, and she is a business administration student at Seoul University. She has just downloaded Tinder, following the advice of a friend. She paid a photographer to create her photo book, and now she is choosing the best pictures to publish on her new Tinder profile. A friend taught her how to use Photoshop, and she's trying to sharpen her skin color. On the advice of her older brother, who lives in San Francisco, she did some research to figure out how to present herself at her best on Tinder and decided to pay the platform to obtain a thirty-minute visibility boost.[65]

This type of strategic agency manifests itself in the visibility enhancing, profile optimization, and rating improvement practices made possible and even encouraged by the platforms themselves. Different actors— individuals, institutions, social formations, and independent content creators, among others—exercise their strategic agency when they plan to gain visibility through the paid promotion of their content (e.g., a post sponsored on Instagram or Facebook by an influencer or institution). Another example of this kind of agency is when an actor entrusts a search engine optimization expert with the optimization of their web page so it can be better indexed in search engines, or a person like Seohyun devotes a lot of attention to the completion of her profile on the Tinder platform, carefully selecting and editing her photos and following the advice of the platform to add more than one photo to her profile to increase its visibility. Examples of this kind of agency can include a young indie music band that decides to invest a few dollars on Facebook ads to boost its visibility, or even a political party, a fashion brand, or a newspaper that plans a series of sponsored posts to enhance popularity of its Facebook accounts or to drive traffic to its website.

Cultural and creative industries are being increasingly platformized:[66] content creators are becoming more and more dependent on platforms, which act as increasingly powerful gatekeepers for the production, circulation, and discovery of their products. Optimization, then, has become a "key business strategy for both platform providers and content producers spurred by the competitive nature of the cultural industries."[67] The American media scholar Jeremy Morris and his colleagues call these strategies "cultural optimization: the process of measuring, engineering, altering, and designing elements. . . . of digital cultural goods . . . to make them more searchable, discoverable, usable, and valuable in both economic and cultural senses."[68]

For example, record labels and musicians have developed various strategies to be more algorithmically recognizable and increase their chances of getting on a Spotify playlist. One musician revealed to us that since Spotify has become his main source of income, he has started to minimize the length of the intro of his songs to immediately grab the listener's attention with a catchy song hook: "People are putting choruses at the beginning of songs now, more so than after a verse. Because the first five seconds, if the listeners hear a chorus, then they're more likely to carry on listening. The reason they're doing that is because then you'll get kept in playlists. So the music itself has been altered to complement the platform in which it's going to get listened to the most on. (Informant H)."[69] This is clearly a strategy, not a tactic, because it reverberates over the long term and is not something that musicians can improvise. Morris calls this kind of practice "sonic optimization," but he also mentions another strategy, "meta-data optimization," which has an impact on music composition:

artists "think of song titles and lyrics not just as signatures of their creative processes, but as keywords that might direct traffic to their content."[70]

Strategies to optimize one's own content or services are spreading across all platforms. We recently studied Airbnb hosts and found that different optimization strategies are widespread among these people.[71] Asking to leave a review is quite a common practice among Airbnb hosts. Yet each host does it in a different way: some ask for it when the guest is checking out, some prefer to opt for an email a few days later, others remind them of the review through a mobile phone message soon after they left the apartment. One host based in Cagliari, Italy, mentioned to us that he welcomes his guests with what he calls a "welcome sheet" (see figure 2.2), in which he explains to his guests how reviews work and, more important, how valuable a five-star rating is for him. To earn and keep the title of superhost (which guarantees major visibility on the platform), he explains to his guests that his listing has to meet certain requirements, one of which is to maintain an overall rating of no less than 4.8. Therefore, since "clients don't know how reviews work—for the average user giving a 4 out of 5 rating is a lot because in school when you got an 8 it was a lot—so I have prepared a welcome sheet in which I explain that for the Airbnb algorithm everything below 5 stars is rated negatively . . . and it really worked."[72]

As we saw in figure 2.1, the differences between strategic and tactical agency are not clear-cut; rather, they are distributed along a continuum: this means that, for example, two forms of strategic algorithmic agency can occupy different positions along this continuum. Let's give two examples. In the first case, a political party invests 100,000 euros in its digital communication campaign, spending 80,000 euros on advertising on Facebook and Instagram for three months. This kind of investment is remarkably different from a small organic meat butcher shop sponsoring itself on Instagram with a single investment of 100 euros, just to see if it will have any effect. Both follow a strategy—investing money in exchange for visibility—but the butcher's agency, compared to that of political candidates, is limited by its low availability of economic resources, time, and technical expertise.

Dear Guest,

we're so pleased that you're going to spend your time at

This apartment is our investment for the future, when we hope we'll be able to move back to Cagliari after many years spent living and working abroad. We use the apartment ourselves when we can, and we've organized it keeping a family need in mind in any aspect. We continue to make improvements and make it better for us and for you.

We're not perfect and we're delegating some of the daily activities to some great family members and friends. If something is not as you expected, we please ask you to inform us immediately and we'll take care of it.

As someone said, feedback is a gift. Only being informed we can improve.

Offering a Five Stars experience it's our main commitment and we did and will do everything we can to ensure that you'll feel our care in any aspect of your stay.

As you might not be aware, platforms like Airbnb and HomeAway require Hosts to be rated no less than Five Stars to keep a certain level of visibility or – for instance – the SuperHost badge. Reviews are of a great tool on those platforms and we want to ensure we get the most out of i

This is why we ask you, again, to keep us informed if something is not as you expected so that we can improve and count on your Five Stars.

Looking forward to hear great stories of your stay and … don't miss an Ichnusa Beer at the ▮▮▮Beach!

airbnb   HomeAway

Phone/Whatsapp/Text

2.2 "Welcome sheet" made by a superhost based in Cagliari, Italy.

TACTICAL ALGORITHMIC AGENCY ALIGNED WITH PLATFORM MORAL ECONOMY

Tommaso is a twenty-four-year-old student at the University of Siena, Italy. He is currently on Erasmus in Amsterdam. He's looking for a girlfriend and a classmate told him to try Tinder. So Tommaso opens a profile on Tinder and starts liking all the female profiles that the algorithm selects for him, thinking that this will increase his chances, but he is not very successful. He doesn't have a lot of money and doesn't want to buy a visibility boost. His Dutch friend tells him that on Tinder, it is better to select "matches" with care, to "educate" the algorithm, and suggested to him a couple of tricks he read in a blog: First, he has to create many different personal profiles with slightly different features, to enlarge the pool of possible partners. Second, when he realizes that the algorithm always proposes the same profiles that fall short of his expectations, he must delete the profile and create a new one, so that the algorithmic evaluation of his interactions will start again from scratch. Tommaso follows his recommendations and eventually gets a few matches, though he doesn't find a girlfriend in Amsterdam.

Tommaso didn't have money to invest in the optimization of his Tinder profile, but he invested his time resources to create and delete many profiles and start over from scratch. He acted in the loopholes left behind by Tinder without violating or questioning its rules. This type of agency differs from the previous one in the different endowment of resources available to individuals or groups. Tactical actions are ephemeral, sporadic, and temporary, and they do not use planned economic investments (as in the case of a sponsored-content campaign).

Moreover, this category includes all those creative practices implemented by actors with limited resources, like shopkeepers, self-entrepreneurs, nonprofit associations, or petty producers who use social media to publicize their brand or activities. Frequently unable to plan long-term investments in promotional campaigns, they use their social capital to improve their online rating or organize free social events where they openly ask to put a "like" on their Facebook page or write a review on their artisanal ice cream shop. Another example of tactical optimization is a common practice spread among Airbnb hosts. Two interviewees revealed to us that they used to delete their listing and redo it from scratch. There are two major reasons for this: first, according to the theories they have developed on how the algorithm works,[73] Airbnb's algorithm rewards new listings with a boost of visibility. Second, when a host receives negative reviews, they believe that it is better to delete the listing and start from scratch, so as not to be penalized by Airbnb's algorithm, which, according to them, makes listings with negative reviews less visible. Although this practice is not explicitly forbidden by Airbnb, it is considered by most hosts a morally ambiguous act.

STRATEGIC ALGORITHMIC AGENCY NOT ALIGNED WITH PLATFORM MORAL ECONOMY

Alfredo del Mazo is the governor of the state of Mexico, and every time he publishes a tweet, he receives thousands of retweets. Researchers[74] have found that he owes his popularity to an army of 60,000 bots. The most curious thing is that practically all the bots integrate images of middle-aged women, and they tweet from the Mexican city of Toluca five times more than from Mexico City, despite having ten times fewer inhabitants.

This form of agency includes all those practices exercised by actors with substantial economic and political capital that can be invested in long-term strategic plans. These plans include a series of actions that may openly or partially violate the terms of service of the platforms, but that, from the point of view of those who carry them out, are considered legitimate means of achieving their aims. Computational propaganda and astroturfing practices that address algorithmic infrastructures are included in this category, as are click farms, click frauds, audience boosting, and fake or paid reviews.

Troll farms are a good example of this type of agency. One of us, together with one of his students, studied troll farms in the Middle East and interviewed employees who work for these companies.[75] The study revealed how these troll farms are often funded by agencies controlled by the government, which seek to create a critical mass of negative and disparaging comments against the regime's political opponents. The trolls work in teams of ten to twelve people, and every day they receive through Telegram the targets that they need to "hit" (i.e., the political enemies that they are asked to attack). They usually have a dozen or so targets a day, and their supervisors set a minimum threshold of tweets and Facebook comments that they must produce by the end of the day. Every day, the trolls "wear the clothes" of at least ten different social media accounts, which they find already open on their computers when they enter their office. From that moment until the end of the day, they must produce at least seventy tweets and three hundred Facebook comments. Trolls represent the human side of computational propaganda, but their activity is augmented by AI. The content produced by them is usually amplified by bots.

It is this complex entanglement of human and nonhuman actors that is responsible for the well-known phenomena of fake engagement and fake trends on social media. In this case, we are talking about strategic agency because behind the troll farms are huge, long-term economic investments by public relations agencies and governments with plenty of economic resources, political power, and strategic vision (the political enemies to target and the political issues to focus on). At the same time, it is a type of agency that is not aligned with the moral economy of platforms but that is shaped by its own morality. Social media are seen as tools for pursuing a specific political agenda, and any means is considered legitimate. In chapter 5, we zoom in on this type of strategic algorithmic agency in the realm of contemporary politics.

Another example is represented by fake reviews on Tripadvisor. Restaurant owners know that the volume and quality of reviews have a direct impact on the visibility of their accounts and sometimes try to game them by paying someone to artificially increase the number of positive reviews. Tripadvisor calls them "paid reviews"—when a business "either knowingly or unwittingly, employs the services of an individual or a company to boost its ranking position on Tripadvisor with positive reviews."[76] Biased positive reviews can also occur when a business offers its customers incentives, such as free meals or discounts, to post reviews. Tripadvisor labeled these efforts as "review boosting."[77] On Tripadvisor and other similar platforms, it is also possible to come across fake negative reviews, paid for by someone who wants to harm a rival. Tripadvisor calls them "biased negative reviews"—when "someone submits a deliberately malicious review about a property in an effort to unfairly lower its ranking position or improperly discredit the property in some way."[78] They also refer to these practices as "review vandalism." We could also call them "reputational wars."

The difference between the strategic and tactical forms of agency that are not aligned with the moral economy of the platforms lies in the fact that the former is more frequently deployed by powerful individuals or institutions like states, governments, political parties, think tanks, and public relations agencies. These actors have great influence and a wide disposition of economic and social capital. The latter instead more frequently comes from those who have little or no power at all, like gig workers, petty producers or shopkeepers, and social movements at the beginnings of their formation. In fact, de Certeau noted that "a tactic is determined by the absence of power."[79]

TACTICAL ALGORITHMIC AGENCY NOT ALIGNED
WITH PLATFORM MORAL ECONOMY

Heathrow Airport, London. Dozens of people who have just disembarked are picking up their phones to book a taxi ride. Meanwhile, an Uber driver shares a message in a Telegram group composed by other Uber drivers like him, and he urges them to disconnect from their apps. The group members log out of Uber, and the price of the ride starts to rise because the Uber algorithm does not find any driver available. When the price is deemed "fair," the only driver left connected to the app to monitor ride pricing gives the others the signal to reconnect. Drivers begin to become available again and accept new rides, but at a higher price than before.

This kind of tactical agency differs from the "aligned" tactical agency not only because it relates to an action not allowed by the ToS, but also because it often requires cooperation among many individuals: so-called surge clubs[80] organized by Uber drivers to raise the price of a taxi ride through the simultaneous logouts of all members of the club, constitute an example of this type of agency. In this category, we include all those practices that involve an individual or collective action aimed at producing a change in the result of algorithmic computation that is favorable to the actors that implement it.

Instagram pods also belong to this category. They are groups of Instagram users who support each other by putting "likes" and leaving comments on the photos of the group members with the aim of increasing the visibility of their content.[81] Neither of these practices is allowed by Uber or Instagram: according to the moral economy of these platforms, users must grow their audiences on their own, in an organic way, relying only on the quality of their services and content that they produce, without the help of anyone else. The moral economy of these platforms is based on the free market of ideas and services, where everyone is in competition with others and where "success" is certified by a favorable position in the ranking produced by the algorithms. Who performs best (a rider who is particularly fast at delivering pizzas; an Airbnb host who takes good care of her customers; an influencer who produces attractive content; or someone who receives many matches on Tinder) will be rewarded by the algorithm with an increase in reputation score. This value system risks producing the famous "Matthew effect"[82] in which the rich get richer and the poor get poorer. Uber workers and Instagram users who organize themselves respond to this logic by trying to gain personal benefits through decidedly rudimentary forms of mutual aid and solidarity. In chapter 3, we will see many of these forms of agency enacted by gig workers.

This tactical use of agency can also be found in the actions of social movements and political activists aimed at increasing the visibility of political causes, which Emiliano Treré[83] has called "algorithmic resistance" and which are explored in depth in chapter 5. These forms of tactical agency may openly result in forms of resistance to platform power that will be described in the following chapters.

IT TAKES TWO TO TANGO: FRAGILE ALLIANCES BETWEEN HUMANS AND ALGORITHMS

These four manifestations of algorithmic agency are intended here only as Weberian ideal types. In reality, we assume that people, social groups, and institutions incessantly move along the two continuums, putting in place different practices that sometimes openly contest the moral economy of the platform, while other times they accept it unconditionally. Individuals, social groups, and institutions may decide several times, during the same day, whether (or when) to team up with algorithms to pursue their own goals or whether (or when) to break this alliance and cheat them. For example, Jude is a twenty-two-year-old student living in London. Every night, he works four hours for Deliveroo, and he has developed a set of tricks to fool Deliveroo's algorithm and earn a handful of extra pounds, but only three hours earlier, upon leaving the university library, he opened Spotify and carefully selected a couple of albums with the intentional aim of "training" its algorithm. Just a few hours before that, on Facebook, he changed his profile picture to Homer from *The Simpsons* because a few days ago, he read something about Facebook's facial recognition software and did not want his face to be used to train Facebook's AI. Jude is also a big fan of BTS, a Korean pop group, and the night before, he stayed up late to participate in the BTS fandom's attempt to get the hashtag of their newly released album into Twitter's "Trending Topics."

People can either ally themselves with platforms' algorithms, even temporarily, and be in solidarity with their decisions (human-nonhuman solidarity), or reject this alliance, partnering with other humans (human-to-human solidarity) to resist them. We call the alliances between humans and algorithms *algorithmic alliances*. These alliances may assume two forms, either (1) *sociomaterial*[84] *alliances* between human actors and algorithmic infrastructures, which happen when humans appropriate and repurpose an algorithm to fulfill a specific need or objective, or (2) *social alliances* between different human actors organized around the need to understand how to better resist the work of algorithms, normally to improve their lives or working conditions.

Every day, people decide in a more or less reflexive manner whether to ally with the computational power of algorithms and accept their ToS and their moral economy, or to break this alliance. It's a dance with algorithms, like a tango. Tango dancers are in constant tension with each other. The grammar of the tango brings them together and apart again and again, resulting in different degrees of intimacy, which are ephemeral and temporary.[85] In the relationship between humans and algorithms, it takes two to tango: humans and algorithms constantly feed each other with data and recommendations, and each is mutually shaped[86] by the acts of the other. Algorithmic alliances in everyday life are like a tango because they can be highly intense, but short-lived; they can break down easily, be transitory, or reinforce each other over time. Unlike the tango, however, the power relationships between algorithms and humans are much less clear and much more asymmetrical.

These alliances have variable durations, as well as forms of resistance, and all of them produce different reconfigurations of the outputs of algorithmic calculation. We know, however, that the relationship between individuals and algorithms is always recursive: algorithms learn quickly from users' gaming attempts and therefore are able to realign themselves. Yet, in the course of this book, we will see that users are also capable of readjusting themselves to face the new challenges posed by algorithms. Thus, we argue, the continuous realignment of algorithmic alliances and resistance gives life to recursive reconfigurations of power balances. We hold that algorithmic systems are to be considered as sociocultural and sociopolitical battlegrounds where platform power and individual agency are continuously renegotiated.

RETHINKING DEBATES ABOUT ALGORITHMIC POWER

This chapter has sketched a multidimensional representation of the nature of the agency still available to individuals coping with algorithms.

Our theoretical framework is able to give an account of the complex entanglements existing between human agency and algorithmic power. Building on both the theory of the moral economy advanced by Thompson and Scott and the concepts of tactical/strategic actions envisioned by de Certeau,[87] we have shown that the relationship between individuals and algorithms does not simply oscillate between two poles (tactical resistance vs. passive acceptance); rather, it can assume a multipolar nature. Our conceptual lens is an attempt to avoid a dystopian narration of the power of algorithms. If Zuboff is right, automated subjects "would allow a fully automated society to run smoothly and frictionlessly,"[88] but the bare existence of these practices is a clear sign of a "glitch in the system" or a proof that the project of automating the subject triggered by the instrumentarian power of surveillance capitalism[89] is not (at least so far) completely successful or frictionless. Moreover, these practices show how users have not yet surrendered to algorithmic rendering of their online behaviors. The tactical exercise of algorithmic agency represents a "threat" to the platform moral economy.

At the same time, we should also be careful not to place undue value on these forms of agency and escape from overly optimistic narrations of users' agency. These practices are often not capable of overturning the balance of power between users and platforms, nor is this their goal (at least most of the time). As Buchanan reminds us in his study of de Certeau as a cultural theorist, tactics "are not in themselves subversive, but they have a symbolic value which is not to be underestimated . . . Tactics are not liberatory in the material sense of the word: the little victories of everyday life do no more (but, also, no less) than disrupt the fatality of the established order."[90]

Perhaps these tactics will not disrupt the platformization of society, but some of them might represent the early stage of more structural forms of solidarity and resistance. The conceptual framework described in this chapter can help us distinguish between forms of agency that aim at pure survival and others that attempt to change the system. In the next chapters, it will guide us in the exploration of the manifestation of algorithmic agency in three fields: gig work, culture, and politics.

3

GAMING THE BOSS

ANKUSH—DELHI

Ankush is a twenty-three-year-old courier working for Swiggy, an Indian online food delivery service platform founded in 2014 in Bengaluru. Since 2021, Swiggy has a presence in more than 500 Indian cities and has a fleet of 200,000 couriers. Ankush lives in Delhi and was forced to give up his education in favor of pursuing work to support his family. This left him with few worthwhile job options. He started working as a courier for Uber Eats, and then for Swiggy in 2019. He had to borrow his father's bike, and then, with his first earnings as a delivery boy, he bought a new one. He told us that in the beginning he used to earn quite well: "Initially, the platform paid us a reasonable income. I used to earn 30K–35K INR per month ($403–$470 per month, n.d.a.). The average salary in India is INR 32,800 per month ($437).[1]" However, as Ankush recalls, those earnings soon began to fall: "Gradually, earnings diminished as there was a surge in delivery jobs. Earlier, there were more orders and fewer couriers, but then couriers outnumbered the area's daily demands. Sometimes I received very few orders that were far from my location, and I could not earn any profit."

With the rise of the pandemic in India, many unemployed workers began working for Swiggy and other Indian online food delivery service platforms, such as Zomato and Uber Eats, and for Ankush, it was becoming increasingly difficult to receive orders. By receiving fewer and fewer orders, Ankush was also at risk of dropping in Swiggy's top courier rankings and getting even fewer orders in the future.

Then Ankush started talking with some of the couriers he met in the streets of Delhi and learned that there are a number of tricks to get more orders and move up in the rankings or take a break from riding twelve hours a day. Ankush discovered that several experienced couriers share video tutorials on YouTube (see figure 3.1), in which they teach the tricks they know or explain their theories on how the algorithms of these apps work.

Among the tricks that Ankush has learned, there is one that he uses often, especially since the orders started to drop. On days when the Uber Eats app doesn't assign him any deliveries, he opens the app from a second smartphone. Ankush enters the app as a customer and places an order, and then he positions himself in front of the restaurant where he placed the order and awaits the order to be automatically assigned to him, the closest courier to the restaurant. The tactics usually works, and he gets the order. As soon as he has picked up the food from the restaurant, he switches to the second smartphone, where he acts as the customer and cancels the order to get the refund from Uber Eats. In this way, Ankush-the-customer gets his money back, and Ankush-the-courier has

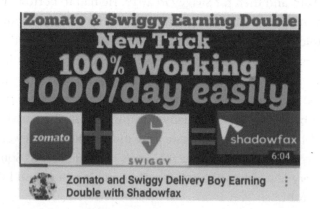

3.1 A still from a typical video tutorial made by experienced Indian couriers and available on YouTube.[2]

undelivered food in his hands. Uber Eats gives couriers the opportunity to donate undelivered food to homeless people, but Ankush keeps the food and secures a free meal.

Ankush told us that he started doing this kind of thing only since the orders started to dwindle. For him, the illegal taking of a free meal from Uber Eats represents a form of "compensation" for the damage that the platform has caused him by hiring new couriers and assigning him fewer orders.

Ankush is just one of the millions of gig economy workers around the world who rely on digital platforms to earn a living and make ends meet.

INTRODUCTION

The story of Ankush was collected for our research project by the Indian media scholar Swati Singh,[3] and it represents a perfect example of tactical algorithmic agency not aligned to the moral economy of the platforms.

In this chapter, we will describe several tactics like this. We will explore the creative tactics, tricks, and individual and collective actions adopted by couriers working for online food delivery service platforms in China, India, Spain, Italy, and Mexico. Yet the same forms of resistance that we observed among the couriers of the online food delivery economy can also be found in many other domains of the gig economy: in fact, a lot of recent research in all domains of platform labor is focusing on how gig workers are devising tactics to partially circumvent, soothe, domesticate, or subvert the power of gig economy platforms.[4]

In the following sections, we will first explore the operational logic underlying online food delivery platforms and the power that they exert on their workers through datafication and gamification, and then we will dive into the stories of agency and resistance among food delivery couriers and drivers.

ONLINE FOOD DELIVERY PLATFORM POWER

Online food delivery companies represent only a small part of the burgeoning gig economy. Gig economy workers make up a large and growing section of the population where short-term, flexible workers are paid upon the completion of tasks (known as "gigs") instead of being paid for the amount of time they work. These tasks are facilitated by digital platforms, fed by data and governed by algorithms that allow them to quickly adapt the supply of workers to fluctuations in demand. The scholarly literature on the nature of the gig economy is now beginning to be substantial, and it is not the focus of this chapter. However, for operational purposes, we must at least provide a definition of this term, which encompasses a whole range of extremely different jobs.

Richard Heeks, a British scholar of digital development, divides it into *digital* (online labor) and *physical* gig economy (location-based service delivery), where "online labor" refers to intangible work delivered online (such as Amazon Mechanical Turk or freelance work platforms like Upwork) and "location-based services" (such as ride-hailing or food delivery) refers to those jobs that are organized digitally but delivered physically.[5] Other scholars propose more subtle definitions of gig work and digital labor.[6] In all these cases, the use of algorithms to govern spatially dispersed workforces is a defining feature of the labor process: algorithms, according to the Italian digital labor scholar Alessandro Gandini, "a) regulate the organization and execution of pre-determined, paid work activities; b) determine the value of such work through opaque calculations; and c) use 'productive metrics,' particularly reputation scores, as tools of control and surveillance."[7]

Gig work is based on algorithmic forms of management. According to David Stark and Ivana Pais, "In contrast to Scientific Management at the turn of the twentieth century, in the algorithmic management of the twenty-first century there are rules but these are not bureaucratic, there are rankings but not ranks, and there is monitoring but it is not disciplinary."[8]

A working paper published by the Organisation for Economic Co-operation and Development (OECD) defines gig economy platforms

as two-sided digital platforms that match workers on one side of the market to customers (final consumers or businesses) on the other side on a per-service (gig) basis. This definition excludes one-sided business to-consumer platforms such as Amazon (trading of goods) and two-sided platforms that do not intermediate labor such as Airbnb (intermediation of accommodation services). As such, gig economy platforms are a subset of the "platform economy" (encompassing any type of one-sided or multi-sided digital platform) and the "sharing economy" (encompassing any type of multi-sided peer-to-peer platform).[9]

Gig work is a world phenomenon. In fact, it is spreading rapidly not only in Western countries, but also in the Global South.[10] A recent projection says that in China, up to 400 million people (half the total workforce) could be employed through gig economy platforms by 2036.[11]

The history of online food delivery is relatively recent. Some companies, such as Just Eat, have been around since 2001, but most have taken off since the second decade of the twenty-first century. China's Ele.me was founded in 2008, as was Zomato in India, while Meituan (China) was founded in 2010, Deliveroo (UK) and DoorDash (US) in 2013, Uber Eats (US) and Swiggy (India) in 2014, and Glovo (Spain) in 2015. App-mediated food delivery is a service that has expanded rapidly worldwide, driven by the global spread of smartphones and the rise of urban, young, credit card–using, middle-class consumers accustomed to shopping online. The rise of these food delivery start-ups has been rapid since their very inception, but it was the COVID-19 pandemic that began in 2020 that turned these services into a global mainstream habit with platform couriers as icons of the global working class during the pandemic.

According to the International Labor Organization, there were 489 active ride-hailing and delivery platforms worldwide in 2020, ten times the number in 2010.[12] The global online food delivery market reached a value of $126.91 billion. New forecasts expect the market to reach $192.16 billion in 2025.[13] These apps are changing the way that we get food on the table, both in the Global North and the Global South. Thousands of delivery workers are whizzing through the streets of global cities, their colorful backpacks slung like cube-shaped bundles over their shoulders. But how do these companies control and govern these huge, constantly moving streams of food without it arriving too cold or spoiled?

FRANK AND ITS SIBLINGS: THE COMPUTATIONAL POWER OF ONLINE FOOD DELIVERY PLATFORMS

Frank is the answer. "Frank" is the name that Deliveroo's designers have given to their complex system of algorithms, what they also call the "dispatch engine." It is constantly calculating and recalculating the "best" combination of couriers to orders using predictions for courier travel time, food preparation time, and other factors. Frank calculates that using machine learning models which are trained on Deliveroo historical data.

The engine of Frank uses complex machine learning algorithms and vast quantities of data to make predictions and decisions about drivers in real time and stack orders based on these decisions. Its developers claimed that "it is even able to change its mind about assignments in response to real-world events, such as travel delays. . . . Frank is a complex system, based on a whole host of machine learning models that we're constantly tweaking and iterating on to improve operational efficiency."[14]

In other words, Frank is the algorithmic interface between the online food delivery company and its courier fleet. Its code contains the mathematical formula that governs the working time and physical efforts of the couriers. Frank is the infrastructure on which Deliveroo's power lies. Every online food delivery company has its own Frank. At Foodora (a Berlin-based food delivery platform), for example, they called their system the "Hurrier."

The power of the new online food delivery platforms, like that of all online platforms, relies on their ability to collect and analyze data. Online food delivery service platforms rely heavily on data and machine learning to increase efficiency and improve workforce management.[15] The power of these platforms is therefore primarily computational. Every action performed by the couriers is broken down, measured, and compared to that of the other couriers on duty. This huge amount of data that Frank and the other algorithms are constantly digesting is used by the platform to predict which courier will be best suited to receive the next incoming order. The physical performance of the couriers is constantly compared against each other in an invisible competition that never ends.

Only the enormous computational power embodied by Frank and his peers allows online food delivery platforms to manage, compare, and rank in real time the thousands of couriers working for them in hundreds of cities around the world.

This computational power has allowed these companies to minimize management costs, but it does not eliminate them altogether. In fact, it would be incorrect to think that algorithms like Frank have simply replaced human management and automated delivery work. Frank and the others are in fact algorithmic infrastructures that augment, not replace, the work of human management. They are like exoskeletons worn by human managers to make their work more efficient. Courier management is not fully automated, but we can think of it as a form of human work augmented by algorithms, or, even better, a task that is done partly by algorithms and partly by humans. In a previous study that one of us did with another Italian researcher on music streaming platforms,[16] we demonstrated the complex entanglement between human and algorithmic labor that exists, to varying degrees, across all online platforms and algorithmic media.

In the case of online food delivery platforms, algorithms automatically assign orders to couriers, but each company has hired humans to constantly oversee this process and control the couriers' movements. These humans, called "dispatchers," work in facilities that provide real-time centralized monitoring, control, and command of the workforce. They have large screens in front of them, on which they can monitor in real time the movements of the couriers who are active at that moment. In front of their eyes, the whole orchestra of couriers is constantly moving, like an urban symphony where the score is written by algorithms and the couriers are the musicians called upon to perform the score to the letter. But the symphony is not entirely directed by algorithms: the dispatchers are codirectors of the orchestra and are able to modify the "score." The affordances of the platform allow the dispatcher to intervene strategically in the assignment of orders, the price of a delivery, and the assignment of a bonus to convince the courier to accept a particularly laborious order: a delivery along a bumpy route or with excessively heavy goods, for instance.

Figure 3.2 shows how dispatchers can intervene in orders. The courier contacted the corporate chat to complain about an order of water bottles totaling 94 kilograms. The dispatcher decided to offer the courier an incentive (bonus) to complete the order:

GLOVO DISPATCHER: I'll add you the bonus

COURIER: No, that's impossible, 94 kg! Is the customer out of her mind?

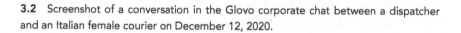

ti aggiungo il bonus

No e impossibile 94 kg

Ma pure la cliente e fuori di testa??

addeso che dobbiamo fare ???
chiamamo il clt per dire che e tropo?

Vuoi venire tu a farlo??

capisco , pero per questo sietti voi con
la macchina

E io capisco che devo caricare e scaricare 47 da
2 litri

se voi riassegnio?

Sirei di si

E francamente stasera troppe cavolate

3.2 Screenshot of a conversation in the Glovo corporate chat between a dispatcher and an Italian female courier on December 12, 2020.

GLOVO DISPATCHER: What do we do now? Do we call the client and tell her it's too much?

COURIER: Do you want to come and do it yourself?

GLOVO DISPATCHER: I understand, but that's why we have you couriers with cars.

COURIER: And I understand that I have to load and unload 47 bottles of 2 litres each.

GLOVO DISPATCHER: Do you want to reassign (the order)?

COURIER: I would say yes.

In this case, the Glovo dispatcher can intervene in the allocation of orders. Each platform has different rules, and not all dispatchers hold the same powers. Other companies worked differently. Foodora, for example, so long as it operated in Italy, outsourced the work of dispatchers to Italian contractors based in Berlin. According to a journalistic investigation by the Italian weekly magazine *L'Espresso*, Foodora dispatchers could intervene in the functioning of the algorithm:

Sometimes we had to fix the algorithm, when the automatic mechanism attributed deliveries in an inefficient way. I followed the riders' routes in detail, I knew the average speed at which they were riding or driving, the entire history of their past rides. If it was raining hard, or there was a very long route to cover, or a dish had to be picked up at a place that we knew was frequently slower than it should be, I could assign "doubles," i.e. double compensation for that single delivery, which was considered more onerous.[17]

Couriers' management is therefore a hybrid job, partly entrusted to algorithms and partly to humans. Thanks to the software at their disposal, dispatchers keep control of entire fleets of couriers, whose perpetual movement generates a constant flow of data, which serves both the algorithm and the dispatcher to make decisions on the allocation of orders. The moving body of couriers is broken down into a flood of data, and the analysis of the data allows the platform to give the courier an overall ranking. This ranking serves as the infrastructure of the gamification mechanisms designed by the platforms to "nudge" couriers to optimize their performance. According to the new media scholars Niels van Doorn and Julie Yujie Chen, the goal of gamification is the same everywhere, for every platform: "to manipulate their flexible labor supply in an agile and cost-effective way—to thereby elicit higher productivity and meet expectations of investors and shareholders."[18]

RANKING SYSTEMS AND GAMIFICATION
Each company has adopted its own ranking system, which is, in part, visible to couriers. Depending on their position in the ranking, the couriers can obtain certain benefits and hope to earn more money. For example,

in India, Zomato ranks delivery workers across four hierarchical levels according to their performance—Diamond, Gold, Silver, and Bronze. Based on their daily experience, our Indian interviewees believe that those who have better ratings get more incentives and better orders in terms of the "fairness" of the price. In China, different online food delivery service platforms set up different ranking systems. These platforms imitate the ranking system of the highly popular Chinese online mobile game *Honor of Kings*.[19] Meituan, one of the most popular online food delivery platforms in China, divides couriers into four levels according to their weekly performance: Bronze, Silver, Gold, and Kings, the highest and most coveted by the couriers. Each level is then subdivided into four sublevels, like Bronze 1, 2, 3, and 4. Ele.me, another popular Chinese online food delivery platform, divides couriers into six levels based on their performance: Bronze, Silver, Gold, Platinum, Diamond, and Kings.

Every platform follows the same basic principle: the higher the level of the courier, the better the chance that they will receive more orders. Higher-level couriers told us that they have experienced higher revenue and better benefits, such as extra orders given by the platform. Chinese high-ranking couriers have a dedicated service hotline and faster appeal efficiency. According to interviewee L, one of the most valuable benefits is that Kings have seven "privilege" orders. So long as couriers have access to the option of "privileged order," they can choose a specific area and they will have priority to select the high-priced orders in that area.

To encourage couriers to take more orders, Chinese online food delivery platforms hold weekly, monthly, and quarterly competitions. So long as the Chinese couriers complete the targets set by the platform, they can get cash or other rewards (such as free electronic bicycles, mobile phones, Bluetooth earbuds, power banks, and other daily supplies) from the platform. For example, according to interviewee I, Meituan used to have summer competitions for Service Platform Couriers (SPCs)[20] in Dongguan. Couriers who earned 63 points during a period of fifteen weeks received an additional cash prize (3,388 yuan, corresponding to about $476).

Mexican online food delivery platforms also follow a similar logic, setting daily, weekly, and monthly targets as performance incentives for couriers, as well as inviting couriers to compete against each other for special prizes.

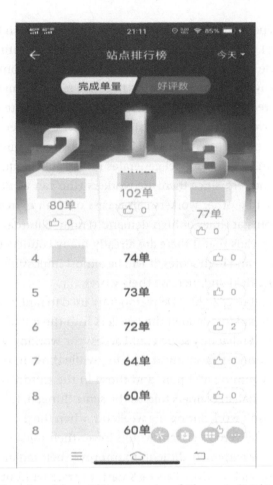

3.3 Daily ranking shared by a Chinese courier of Ele.me in Knights League (KL)'s WeChat group, November 18, 2020.[21]

Online food delivery companies in Italy and Spain similarly adopt gamification mechanisms. To access working hours, Glovo requires that couriers reserve them in advance. Access to these working hours is conditioned by the courier's level of excellence (ranging from 1 to 100), which is mainly determined by a black-boxed mix of the following factors: seniority, efficiency, hours worked during high-demand hours, and customer ratings. This system of "excellence" represents, for all the couriers interviewed, the main source of emotional stress. At the same time, workers feel that they are in a situation of defenselessness (especially in the face of dependence on customer ratings, against which they have no right of reply). David, a Spanish courier,[22] got a poor rating and asked Glovo why the customer had rated him negatively. The company refused to give him any information, and the negative review from the customer immediately reduced his level of excellence. A higher score allows earlier access to the work shift booking system: those with higher scores have a better chance of finding more available shifts or more shifts at times of high food demand. This model generates remarkable inequalities within the fleets of couriers, since there are workers who can work up to twelve hours a day while other workers can access only an extremely reduced number of hours at times of high demand (Friday, Saturday, and Sunday nights). This means that if there are already many couriers in a city with long experience and high scores, it will be almost impossible for newcomers to enter the field and increase their scores.

Until November 2, 2020,[23] Deliveroo Italy used to rank couriers as well. The Deliveroo ranking divided the couriers into three brackets. Couriers with the highest reliability score could access the working hours booking system every Monday morning at 11 a.m., while those in the second tier could do so beginning at 3 p.m. and those in the third tier only from 5 p.m. on. Many Italian couriers told us the same thing: if, for some reason, someone couldn't work during the weekend, when the demand for orders is at its peak, Deliveroo's system would lower their reliability score, and then it would be extremely difficult to improve their rankings.

Other apps, such as Uber Eats or Stuart, operate without a rating and excellence system and allow their workers to log on at any time and start taking orders. This offers a higher level of autonomy to couriers. Yet, since couriers' income depends only on the number of packages delivered and not on the hours they worked, they usually tend to remain logged in to the platform for as long as possible, even when the frequency of orders is meager. Although these apps have apparently abandoned the idea of taking statistics, many couriers have told us that they suspect that orders are awarded on the basis of rankings that are invisible to them, even though they cannot prove it. For instance, according to his experience as a courier and a member of a Spanish trade union, Pepe is convinced that some online food delivery platforms give preferential treatment to certain couriers who publicly support their ethics or are extremely compliant with their rules. In addition to these ranking systems, platforms use other systems to incentivize couriers to work harder and compete against each other. Online food delivery platforms make frequent use of algorithmic nudges to influence their couriers. These nudges include surge pricing predictions, notification pushes, and weekly performance reports. Surge pricing, in particular, is an algorithmic-oriented system that uses price adjustments as financial incentives to redistribute the workforce in a territory,[24] and it is widely used by ride-hailing platforms too.

In Italy, for example, as of November 2, 2020, Deliveroo modified the maps of the cities where couriers work and divided the urban areas into hexagons. The higher the figure within the hexagon, the more orders available in that area. The app sends notifications to couriers' mobile phones to let them know that orders coming from certain area of the city are expected to surge in the next few minutes. Uber Eats adopted similar heat maps too, as seen in figure 3.4.

The computational power of online food delivery platforms allows them to govern a dynamic workforce and respond in real time to changes in demand. This power severely limits the autonomy of couriers, binding their actions to the technological affordances and legal rules of the platforms.

The increase of computational resources at the disposal of gig economy companies in such markets as online food delivery has shifted the balance between capital and labor in favor of the former. Yet, as we will see in the following sections of this chapter, this relationship, though strongly asymmetrical, rests on an unstable equilibrium: we will show how couriers have developed different tactics to resist the computational power of platforms, exploiting their loopholes and eventually "gaming" their (algorithmic) bosses.

GIG WORKERS STRIKE BACK: RESISTING PLATFORM POWER

Food delivery couriers perform a highly fragmented job: each one darts down the street locked in their own bubble, constantly bouncing like a pinball from restaurants to customers' homes. Couriers do not share the same workspace, as factory workers do. This spatial fragmentation slows the establishment of collective consciousness among them. The

3.4 Areas of higher demand on Uber Eats (1.3x) in Turin, Italy, in March 2021.

Italian social movement scholar Donatella della Porta and her colleagues showed how the work of food delivery couriers exhibits at least five types of fragmentation: legal (they are self-employed), technological (they are individually governed by an algorithm), organizational (piece-rate work), spatial (they are dispersed in urban space), and social (high ethnic heterogeneity).[25] Yet workers continue to organize collective action[26] and "mobilize against the odds," as della Porta et al. rightly point out.[27]

In fact, within a few years from the founding of the first gig economy start-ups, gig workers realized the precariousness and lack of rights in their new jobs. Soon, protests, strikes, and riots began to break out everywhere. Deliveroo, one of the world's most famous food delivery companies, was targeted by protesters for the first time in 2016, only three years after its founding. On August 10, 2016, after testing a new payment model, which the company claimed had been a success, Deliveroo UK began paying couriers a fixed price per delivery of £3.75 rather than the current hourly rate of £7 per hour plus £1 per delivery. This has left some of Deliveroo's UK couriers unhappy enough to hold something akin to a strike. Outside the Deliveroo headquarters in London, more than 100 self-employed workers gathered calling for the company to keep the hourly rate.[28]

NOT ONLY RIOTS: TACTICS OF EVERYDAY RESISTANCE

However, these expressions of anger and protest against the ominous consequences of the rise of platform capitalism should not be seen as an entirely new phenomenon. The human geographers Jamie Peck and Rachel Phillips critique the exaggerated claims to the novelty, and indeed the revolutionary significance, of the concept of platform capitalism through a Braudelian approach that allows them to situate platform capitalism within the long duration of industrial capitalism: "The Braudelian schema requires that platform capitalism is situated, both historically and geographically, in this case both as a distinctive conjunctural moment and as an epiphenomenon of variegated and globalizing processes of financialization and neoliberalization."[29] Core features of platform capitalism like monopolization and concentration, according to them, are recurrent tendencies of the capitalistic system. Then, if platform capitalism is merely the mask that capitalist accumulation has taken on in the digital age, platform workers' contempt and dissent against it are also merely the latest stage in a long history of workers' protests that always operated to subvert, alter, or take over the processes of production. There is a large body of literature on how this operated during industrial capitalism[30] and during the early stages of the web economy.[31]

Riots, unregulated strikes, and protests are a prosecution of those past workers' struggles and represent only the most obvious manifestations of the dissent and discontent that has animated gig workers in recent years. For city traffic blocks, couriers' pickets in front of a platform's headquarters and riots explode when anger and discontent reach unbearable peaks when workers see no other way to get by and make ends meet. Yet these forms of public resistance to the power of the platforms are only the tip of the iceberg, the most visible form of dissent that runs like an underground river through the fabric of gig workers' daily lives. Most gig workers do not want to destroy the work that lets them earn a living, at least somewhat. They just want to improve their conditions. When gig workers do not take to the streets to protest, that does not mean that they are happy with their jobs—they are simply busy surviving, trying in various ways to make the most of the work. The ways in which they try to survive in many cases lead them to perform actions considered illegitimate or immoral by the platforms. These actions are precisely the subject of this section and, inspired by the work of James Scott, we understand them as contemporary forms of "everyday resistance."

Everyday resistance, a theoretical concept introduced by Scott,[32] describes all forms of dissent that are not as dramatic and visible as riots and revolutions. Everyday resistance behaviors of subaltern groups (for example, foot-dragging, escape, sarcasm, passivity, laziness, misunderstandings, disloyalty, slander, avoidance, or theft) are typically hidden or disguised. As highlighted by the resistance scholars Stellan Vinthagen and Anna Johansson, these activities are tactics that "exploited people use in order to both survive and undermine repressive domination; especially in contexts when rebellion is too risky."[33]

Wherever algorithms are employed to manage workflows and workers' actions, we discovered tactics invented by the workers themselves to partially and temporarily circumvent or subvert platforms' computational

power. The creativity of gig workers in crafting tricks and inventing practices to game the algorithmic systems that govern them is astounding.

The delivery drivers working for Amazon are driven by an algorithm
that constantly monitors them and optimizes delivery times. They are
under so much pressure that some of them have admitted to urinating in a
bottle while driving to save time. To alleviate this pressure, even partially,
the drivers have developed very creative tactics.[34] Even people who work
in Amazon's warehouses are subject to enormous scrutiny and control by
both human supervisors and automated systems. Every picker (the person who searches for goods on the shelves of the Amazon warehouses) is
equipped with a scanner. The scanner is a powerful surveillance tool that
records the picker's productivity rate—displaying it on its interface—and
the time between the scan of one object and the next one (called Time
Off Task, or TOT).[35] If the TOT of a picker exceeds fifteen minutes or
their rate falls below the prescribed speed for the day, they will get a visit
from a manager or a write-up. "Rates are used as Damocles' sword," Char-
lie said. "You can be king, but there's a blade hanging above your head
held by a thin hair."[36] At the end of each daily shift, managers publish a
ranking of each worker's productivity, and the most productive get small
rewards like Amazon mugs or T-shirts.

According to the American journalist and writer Sam Adler-Bell, who did
research on resistance tactics inside Amazon warehouses, online Amazon
worker forums are full of descriptions of strategies for artificially boosting
rates.[37] As reported by Adler-Bell, "One worker discovered that managers
were basing his productivity numbers on how quickly he started work after
a break. By leaving a count loaded in his scanner, he could trick the computer into thinking he had resumed work with a flurry of activity. Others
boost their count by rapidly scanning several bins of small items."[38]

Like other similar resistance practices employed by workers of the
Industrial Age, these are only temporary remedies to the pressure exerted
on them by the complex entanglement of human and algorithmic workforce management. At least, they allow workers to catch their breath during grueling shifts, to escape the surveillance of the algorithmic boss for a
moment, to bring home a few extra dollars at the end of the day, or simply to put a stick in the wheels of platform capitalism and slow its race—if
only for a few seconds. Algorithms, as Fabian Ferrari and Mark Graham

argue, "do not have hegemonic outcomes, and they do not entirely strip away agency from platform workers."[39]

EVERYDAY ALGORITHMIC ALLIANCES AND RESISTANCE AMONG FOOD DELIVERY COURIERS AND DRIVERS

What drives couriers around the world to break the rules of the platforms they work for and risk being banned by them? What motivates them to team up with each other to beat Frank?

From Mexico to India, from Spain to Italy, couriers gave us similar answers: they want to improve their living conditions and have a decent job that allows them to support their families or their studies. They have no interest in ripping off the multinational companies they work for, but most of them recognize that if they had always respected the rules of the game, they would not have been able to survive.

There are many differences between the couriers we interviewed in India, Italy, Spain, Mexico, and China: each of them holds different cultural and social capital and belongs to manifold social classes. Education levels among couriers are highly variable, not only from country to country, but also within the same city. Yet, despite the considerable differences in class, age, ethnicity, and cultural capital, the tactics and strategies developed by couriers to curb the power of platforms are strikingly similar. Faced with similar exploitative conditions, couriers from all over the world reacted in a similar way, regardless of their geographical or cultural differences. In all cases, we noted the emergence of patterns of resistance based on similar practices, which can be divided into two macrogroups: *individual* and *collective* tactics and strategies. The former are practices employed by individual couriers to improve their personal condition, even to the detriment of their colleagues. The latter are always the result of coordinated actions, requiring time and cooperation among colleagues to be effective and aiming to improve the collective working conditions. Both sets of practices, however, are learned, discussed, and refined in the private online chat groups created by couriers to talk about their work, and to which we will devote specific attention later in this chapter.

However, since the relationship between couriers and platforms is symbiotic and dynamic, our findings can only crystallize specific moments.

As soon as platforms change their rules, tariffs, and algorithms, couriers adapt accordingly and develop new practices, which quickly circulate in the private chat rooms. At the same time, as soon as a platform discovers an illegal practice, it introduces new changes to prevent couriers from continuing to benefit from it. The relationship between couriers and platforms can be represented as a constant cat-and-mouse game, or as a tango in which one of the two dancers is much more dominant than the other.

In the following discussion, we will describe the individual and collective tactics that occur most often among online food delivery couriers. Some tactics have been omitted at the explicit request of our interviewees, either because they are practices still unknown to the platforms or because of the fear of being discovered by the platforms themselves.[40]

INDIVIDUAL TACTICS AND STRATEGIES BETWEEN RESISTANCE AND OPTIMIZATION

Some couriers refuse an order, while others keep a diary of their movements and earnings. Some couriers take shortcuts to save time, some eat undelivered orders, some arrange with a restaurant to pretend to deliver a pizza, some resort to bots to book work shifts, and some invent multiple identities to have a better chance of receiving an order. The galaxy of individual tactics to make ends meet is vast and ever-changing. From this galaxy of practices, we have chosen nine as manifestations of algorithmic agency (a concept discussed in chapter 2) exercised by couriers.

DIARIES AS "TECHNOLOGIES OF THE SELF"

Some couriers keep diaries, either in paper notebooks or on their smartphones (see figure 3.5). For example, Bruna, one of the few female couriers working in Florence, Italy, has been keeping a handwritten diary of all her deliveries for two years. She takes note of the price, the distance, and the location of each delivery she makes, and before accepting a new order, she checks whether the distance estimated by the food delivery app corresponds to that provided by Google Maps. Sometimes she discovered significant differences between the two estimates. Bruna uses the notebook to keep track of her earnings, to monitor her work, and to

3.5 Screenshot of an electronic note taken by a courier working in the hinterland of Milan, Italy.[42]

understand the days and time slots in which she receives more orders. She looks at these data in search for emerging patterns, such as areas where the algorithm sends her more often to make deliveries.

Through these diaries, couriers try to regain a share of control over their work. Instead of being passively datafied, they can reappropriate information to accomplish, by their own limited means, what algorithms do in the blink of an eye: gather and analyze data. They employ diaries as tools to optimize their work and improve their performance. Here, the diaries function as "technologies of the self"[41] because they allow workers to improve their knowledge about themselves.

We can thus understand this practice as a form of algorithmic agency that can be both strategic and tactical (depending on the resources and knowledge available for the analysis of their own data). Tactics like this tend to increase worker productivity, and thus generate greater value for companies. We cannot properly consider them forms of resistance because they are aligned with the moral economy of platforms, although, as in this case, they try to resist passive datafication of one's work and regain some form of control over it.

AUTOMATIC BOOKING OF WORKING SHIFTS

The practice of automatic booking of shifts generates a high volume of conversations in couriers' private online chats and is also one of the most hated among them. Bots are subscription-based apps that can cost as much as 30–50 euros per month. They are effective only in cases where couriers work on platforms that adopt ranking systems that determine access to shifts. In this kind of platforms, such as Just Eat or Deliveroo,[43] couriers have to book the hours made available by the system for each area in which a given city is divided. As soon as the system makes new shifts available, hundreds of couriers are ready with their phones in their hands to book them. If they arrive late, they risk not finding any more available hours, which can have a negative impact on their ranking (not to mention their income).

Bots help eliminate this risk: the app installed on the smartphone is always active in the background, and as soon as a shift becomes available, the bot automatically books it for the courier (see figure 3.6).

Couriers who use bots accumulate more hours than those who book their hours manually. This practice is highly criticized by couriers, though, as it is considered unfair competition.

In these cases, we can define the agency of those using bots as a form of strategic algorithmic agency, which uses economic resources (the monthly subscription to the automatic booking service) and long-term computational resources to optimize the performance of the couriers, in line with the moral economy of the platform, which does not prohibit the use of bots and nudges couriers to compete by all possible means. As with the case of keeping diaries, this tactic does not challenge the moral economy of the platforms; rather, it tends to optimize it, seeking to make the best use of bots to better survive competition with other workers.

WORKING FOR MULTIPLE PLATFORMS

Zizheng, a Chinese courier, usually uses multiple mobile phones to register on different platforms to ensure that he is available for work on all of them at the same time. Working for several platforms at the same time is an extremely popular practice and the subject of several studies.[44] It is not banned by platforms in itself, although some countries such as China do

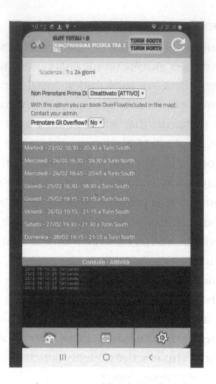

3.6 Screenshot of an app for automated booking of working shifts, running on the smartphone of a courier in Turin, Italy.

forbid this activity. Platforms allow couriers to work for more than one food delivery company only because banning that could be used by European trade unions to prove that couriers are not autonomous workers, as the platforms would like to present them as being. Even though this practice is not banned, platforms highly discourage it, trying to retain their "partners" (as they call the workers) through complex mechanisms of gamification, incentives, and bonuses. This kind of practice lies somewhere between the moral economy of the platforms and that of the couriers: although the platforms do not forbid it, it is strongly discouraged. Workers who adopt it say that they do so unwillingly, and they justify the potential harm to the customer by blaming the platforms for lowering rates too much and pushing them to find viable alternatives to increase their earnings. It is a tactical form of agency, with no long-term strategy. At times, it is adopted as a safety net when a courier is not receiving enough orders.

MULTIPLE ACCOUNTS ON THE SAME PLATFORM

Another widespread practice is to have several accounts open on the same platform to increase the chances of receiving an order. This practice is also highly criticized by most couriers, as it is considered unfair competition. A courier who runs several accounts must have several mobile phones, on each of which the courier has registered a different account under the names of friends, brothers, wives, grandparents, or other relatives who have agreed to provide their personal data. Such a person is working, literally, "for three."

This practice is considered illegal by the platforms and strongly discouraged: if a platform discovers that a courier is using different accounts at the same time, it can ban her for months or even altogether, without providing any explanation for its decision. The Spanish food delivery company Glovo, for example, has introduced facial recognition software to verify the identity of each courier at the beginning of a shift. The platform, therefore, imposes a deeply invasive surveillance on its employees, and yet one courier from Naples, in a private chat on WhatsApp, explained how to "cheat" the software: "I scan my cousin's ID. I put my picture in place of his, I scan it, and I send it to Glovo. Then I open an account in my cousin's name. I have three accounts with the same face. You can counterfeit anything!" This tactic also tends to optimize worker productivity at the expense of others. It is, however, only partially aligned with the moral economy of platforms because while it accepts the ethics of competition, it does not accept the ethics of surveillance imposed by them through facial recognition software and identity monitoring.

ORDER REFUSALS

Delivering a Peking duck in the suburbs of a Chinese metropolis can be a feat worthy of the protagonists of *The Fast and the Furious*: traffic, construction, police, no-go lanes, bad roads, and unknown street names can slow the delivery considerably. We discovered that in China, experienced couriers used to reject orders from a certain region, indirectly forcing the platforms to increase the delivery price or provide extra tips. These couriers know that orders in some areas are more difficult to complete for reasons that can range from the absence of an elevator in high buildings, which forces the courier to climb multiple stairs, to the difficulty of completing an order in an area that is too big or complex. When they encounter orders from these areas, couriers do not accept them (e.g., they can directly reject orders dispatched by the system or choose to take a break by logging out of the system) because they consider the price offered by the platform to complete the delivery as too low.

In an Italian courier chat, for example, one worker suggested the following tactic to group members, based on rejecting unprofitable orders: "We need to *educate* the algorithm at our own pace: since every time you reject an order its price is recalculated by increasing a percentage of compensation, regroup in the same spot and reject orders until you see the most suitable compensation appear" (emphasis mine). As noted by Francesco Bonifacio, who produced an impressively rich ethnography of Italian food delivery workers, "Learning to select which deliveries to accept and which to reject is a way of gaining control over the temporal rhythms of the work practice."[45] Through this practice, couriers exercise a form of tactical agency aligned with the moral economy of the platform.

DO NOT FOLLOW THE PLATFORM ROUTE

Most food delivery platforms do not offer their couriers hourly rates but rather pay them a dynamic rate for every order they deliver. This logic, together with gamification mechanisms, puts a lot of pressure on workers, who try to speed up delivery times to be available to receive a new order as soon as possible. The faster couriers are, the more they earn. How can delivery times be shortened? They can ride harder, but they can also use shortcuts.

Experienced couriers usually take various shortcuts based on their own urban experience to save the delivery time for each order (even though some of them are obvious violations of traffic rules). In figure 3.7, we can see a conversation between Chinese couriers in a private chat room, where they exchange information on how to deliver faster.

Chinese couriers do not follow the route suggested by the algorithm; instead they choose to create their own paths through urban traffic, just as the pedestrians described by de Certeau[46] used to create their own paths in the interstices of urban spaces to connect isolated areas. This practice

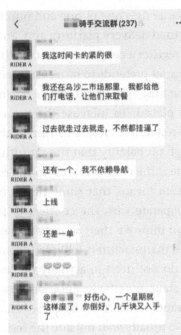

3.7 In a private WeChat group set up by couriers, courier A is sharing his experience of how to complete more orders with others, "Because the delivery time is tight, I don't follow the routes set by the platforms." courier A says. Courier B thanks A for sharing, and courier C compliments A for completing so many orders.

represents a form of tactical agency, but it does not compete with the moral economy of the platform. Rather, it tends to optimize the urban performance of the couriers without submitting their bodies to the automated global positioning system (GPS) guidance of the platform.[47]

SHUADAN

In China and India, some couriers have registered on food delivery platforms as both couriers and customers. When orders decrease or when a courier wants to increase her ranking, they sometimes resort to a tactic that Chinese couriers call *Shuadan*, a Chinese verb that means "to create fake orders artificially." Two types of *Shuadan* exist: courier-led and restaurant-led. Courier-led *Shuadan* means that a courier uses two or more mobile phones to act as the customer and the courier at the same time.

In the first step, the courier uses one phone and number to register a customer account on one food delivery platform app, and then uses another phone and number to register a courier account on the same platform. In the second step, the courier pretends to be a customer, places an order on the first phone, and then use his second phone to immediately grab the order that he has just placed. To increase the success rate of *Shuadan*, the courier chooses a restaurant closest to him with no other couriers nearby, so there is a high probability that the algorithm will dispatch the order to him because he is the courier closest to the restaurant.

Restaurant-led *Shuadan* means that some newly opened restaurants/food-manufacturers cooperate with the couriers to *Shuadan* in order to receive more orders and improve their scores and rankings on the platforms. Restaurants/food manufacturers will play the role of fake customers, place orders (they do not need to actually prepare the food) on the platforms from their restaurants (the location of these fake customers are usually near the restaurants, which is set by these restaurants in advance), and ask the couriers who usually wait outside the restaurants to "pick up" the orders and to push the "Delivered" button directly on the platforms' apps. By doing so, the courier can get more orders and obtain the delivery fees for doing nothing.

We can understand this tactic as a manifestation of algorithmic tactical agency not aligned with the moral economy of the platform because it attempts to hack the order allocation system and partially take control of it.

DELIVERING OUTSIDE THE PLATFORM

In India, while working for online food delivery platforms, couriers find other ways to earn money at the same time. In Delhi, sometimes customers, in addition to the order placed on the platform, ask couriers to bring grocery products from the shop nearest to them. For this "double" delivery, the couriers receive an extra payment from the customers. On other occasions, they use the delivery route to carry out other small jobs or deliver other types of goods to regular customers who contact them directly on the phone, bypassing the platform altogether.

ORDER STEALING

Stealing an order is a common practice in all the geographical contexts that we studied, but only under specific conditions. In such cases, couriers accept an order and heads toward the restaurant to pick it up. At the restaurant, they pick up the food, but, instead of confirming on the app that they have done so, they reject the order. At this point, the order returns to the system and is assigned to another courier, but when the second courier arrives at the restaurant asking for the order that they have been assigned to, they discover that the order has already been picked up. Manuel, a Spanish courier, says that this is a practice that almost all the couriers have experimented with at least once in their life. Yet couriers from China, Italy, Mexico, and India told us that they only decide to steal an order when they are starving or when they feel resentment toward the platform. They understand this as a form of mild retaliation against the platform's exploitation. This practice is morally frowned upon by couriers in the Western world, but in different contexts, such as India, it is considered legitimate in cases where few orders arrive and couriers are profoundly poor and hungry. It is clearly a form of tactical agency not aligned with the moral economy of the platform, which does not include the possibility of appropriating the order.

COLLECTIVE TACTICS AND STRATEGIES BETWEEN RESISTANCE AND OPTIMIZATION

In addition to adopting individual practices of resistance or optimization of their own performance (more or less aligned with the moral economy of platforms) couriers have developed collective practices, which require coordinated actions to achieve common goals. These are forms of cooperation and solidarity that challenge the individualistic logic encoded in the algorithms that govern commercial platforms. These practices represent forms of agency that are not aligned with the moral economy of platforms because they reject the ethics of competition at all costs inscribed in the food delivery apps and contrast it with the value of solidarity.

SOLIDARITY LOG OUT

Couriers work as hard as possible, and thus stay connected to the app as much as possible, to improve their rankings. Better rankings allow them to receive even more benefits, such as being able to book upcoming shifts before others or being able to get greater incentives. In this labor context, organized like a medieval tournament, we would not expect to find supportive practices among couriers, as they are pushed to be concerned only with maximizing their score. Yet in several countries, we have observed various examples of forms of solidarity among couriers aimed at redistributing opportunities to increase individual scores. Couriers who have already reached their daily goals log out of their apps to allow those who have not yet done so to receive more orders, and they coordinate their logouts through private WhatsApp or WeChat groups. We observed this practice in Mexico and China. In Mexico, couriers help each other so that everyone can receive enough orders and thus maintain their positions in the platform ranking. By collectively disconnecting from the app, they create an artificial increase in courier demand, increasing the chances that couriers who are still active will receive more orders.

COORDINATED ORDER REFUSAL

Coordinated order refusal is similar to individual refusal of an order, but it is carried out simultaneously by many couriers and it is orchestrated via private online chat groups such as WhatsApp and Telegram so it can cause a greater impact. The objectives of these collective refusals can vary. For example, in 2021, the US drivers of the DoorDash app launched the hashtag #DeclineNow: more than 26,000 DoorDash drivers belonging to the same Facebook private group collectively declined cheap orders to prod the platform to raise the price of the deliveries:[48] in the absence of traditional collective bargaining intermediaries, the #DeclineNow strategy was aimed at boosting the base pay for drivers by collectively gaming the DoorDash algorithm by exercising their right to choose which deliveries to accept on the platform.[49]

In Livorno, Italy, in 2020, a small group of Deliveroo couriers decided to refuse any order coming from the local McDonald's as a sign of protest against its slow delivery times. This can be considered as an example of a strategic algorithmic agency not aligned to the moral economy of the platform.

MUTUAL EXCHANGE OF WORKING HOURS

Online food delivery apps that adopted automated systems for shift booking do not allow couriers to swap shifts. There is no boss to ask the favor of working a different shift and being replaced by another colleague. Yet Italian couriers working for Just Eat and Deliveroo invented creative methods to reallocate working hours according to mutualistic principles, as can be seen in figure 3.8.

October 5, 2020:
Courier A:
Who wants (a shift)?
   - Central zone (of Florence)
   - On Just Eat platform
   - Tonight, from 8.30 pm to 11 pm
Courier B—Are there any shifts available on Wednesday or Thursday?
Courier C—I also look for evening shift in Novoli area (outskirts of Florence)

In private online chats, couriers look for and offer hours of work. If couriers booked a time slot that they no longer need, they offer it to others who need it the most. When they find another courier interested in their working hours, they agree to release their shift on the app and give it to the other courier. But how is this possible if the app does not allow the intentional transfer of hours between couriers? As soon as a shift is freed up by a courier, the app makes it available again, and the shift could be booked by anyone. So, in the case that courier A wants to free up a shift because they no longer needs it, they would first publicize their will on a private online group they belong to. Then, for example, courier B replies by saying that they need that shift. Courier A then informs courier B via chat that they will release her shift at time x. Courier B, then, is ready, smartphone in hand, to click on the shift as soon as the app makes it available again. This informal agreement between the two couriers minimizes the chances of that shift being taken by someone else. This tactic is also the result of a cooperation between two or more couriers mediated by private online chats. It can be understood as a form of tactical agency

3.8 Screenshot of a conversation between two couriers looking for or offering working hours on October 5, 2020.

that is not aligned with the moral economy of the food delivery companies, which instead designed a platform that encourages everyone to compete with everyone else to grab work shifts.

SABOTAGE

During the riots on November 4–5, 2020, in Milan (see figure 3.9), couriers who had taken to the streets to protest against the new labor contract organized themselves to sabotage the various platforms' food delivery

3.9 Food delivery workers' riot in Milan on October 4, 2020. Courtesy of an anonymous courier.

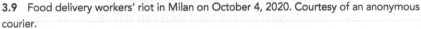

systems. The protest was organized in the previous days on private online chat groups, and hundreds of couriers agreed to log out from the apps simultaneously. During the riots, couriers protesting in the street confronted those who were making deliveries, preventing them from completing their work. The aim was to minimize deliveries, which would damage food delivery companies.

Coordinated, collective disconnection actions like the one in Milan are quite common.[50] We could also liken these acts of sabotage to the forms of Luddism that ran through the history of the Industrial Revolution. In this case, however, there are no machines to destroy. The "machine" is a complex algorithmic system. The new Luddites, like their forebears, do not break the machine because they are against technological progress—they are breaking or damaging the "machine" because it is worsening their working conditions, just like the sabotage of the new threshing machines done by the Malaysian peasants described by Scott.[51] Their aim is to interrupt, even temporarily, the cycle of accumulation of platform capital, just as the Luddites did with factory machinery. The social critic Gavin Mueller described the motives of nineteenth-century Luddites to break machinery, and these motives are remarkably similar to those of today's couriers to "break" platform apps by ripping off their algorithms: "Physically separated and without established organizations, they often related to bosses according to individualized contracts, and so it was impossible for them to engage in the kind of militancy we associate with trade unions made up of mass workers."[52]

These acts of sabotage are examples of strategic algorithmic agency that are not aligned to the moral economy of the platform.

CONTESTING PLATFORM AFFORDANCES

As Winner's classic study argued, all artifacts have politics.[53] Similarly, food delivery apps have politics too. Their politics is to discourage any exchange of information between their workers, so they have not equipped their apps with a chat room in which couriers can talk to each other, but have instead decided to provide a "customer service" chat room that couriers (conceived as customers of the platform) can rely on in case of emergency. These apps afford a specific form of governance of the workforce, enforced by algorithms and aimed at building a vertical, asymmetrical and individual relationship with workers.[54] They allow workers to find goods or food to be delivered, but they do not treat all workers equally. The algorithms favor those workers who are more willing to work weekends or longer shifts or who have a higher level of physical dexterity or more time to spend waiting for an order.[55] Workers cannot benefit equally from the affordances of food delivery apps, not only because of the structural inequality built into these apps through the design of a gamified environment, but also because, as the affordance scholar Jennie Davis reminds us, the mechanisms of affordances are inseparable from the social and structural conditions in which they are enjoyed by users.[56]

However, platform workers are not passive users of these platforms and have started to contest platform affordances. Tech-savvy couriers have created unauthorized apps that enable functions not allowed by the corporate apps, illegally forcing their technological boundaries. For example, in Indonesia, where most couriers work for the Gojek platform, the researcher Rida Qadri discovered that "over the last six years, a burgeoning underground market for unauthorized, third-party Gojek apps has emerged. Named after a child-like spirit in Indonesian folklore that helps his human master earn money by stealing, each *tuyul* (non-authorized, n.d.a.) app responds to specific needs of drivers to help make their jobs less miserable."[57]

Since Gojek came into existence in 2015, more tech-savvy drivers began helping other colleagues solve technical problems created by the app. Eventually, small technical support tasks such as factory resets, reboots, and GPS fixes led to a variety of unofficial applications for different drivers' needs. Some of these illegal innovations introduced by Indonesian couriers were even copied by Gojek itself. In so doing, Gojek improved its service by appropriating the widespread innovation emerging "from below" and eventually saving research and development resources. Yet Gojek drivers were able to force the platform to revise its affordances, and therefore to improve their own working conditions. This development of new technological affordances from below shows a form of strategic agency only marginally aligned with the moral economy of platforms.

PRIVATE ONLINE GROUPS AS INFRASTRUCTURES OF LEARNING, RESISTANCE, AND SOLIDARITY

Just like policemen have walkie-talkies, we have these Telegram groups. Whatever problems we want to discuss, we do it there. We start a movement from there.
—Salauddin, an Indian ride-hailing union leader in Hyderabad, India[58]

By far, the most popular and effective of all the abovementioned tactics is to create private online chat groups on platforms such as WhatsApp, Telegram, WeChat, and Facebook to learn the tricks of the trade, exchange information, learn how the algorithm works, organize collective actions, and provide mutual support. We could also call it an *infrastructural tactic*, meaning a tactic that provides an infrastructure for the exchange of all the other tactics. It is through these private online groups, invisible to the platforms, that couriers discover new tactics and learn how to put them into practice. Without the existence of these groups, it would be impossible to coordinate collective actions or learn new tricks.

In every city, there are dozens of private online groups created by couriers working for a specific platform. Besides these groups, there are larger, national groups, founded by various associations of couriers, emerging or traditional trade unions (both right- and left-wing), which try to channel the discontent of couriers toward more traditional forms of intermediation and protest. Every courier participates in dozens of these groups, both local and national, and over time, they also create smaller groups of five to ten participants at most, to which they invite only colleagues with whom they have developed a friendship and solidarity.

The ties developed in these smaller chat groups are stronger than those built in the large, private groups. One Chinese courier expresses a clear example of this kind of solidarity:

The four of us met in this WeChat group ("Changan"). We often chatted in this group and got along very well, hence we decided to set up a small WeChat group with only four of us. We chat more frequently in groups, and we often meet offline. We share our experience and various tactics and information in time to truly help each other. Once I accidentally ran into a car during the delivery of one order, I asked for help in the group. The three of them quickly arrived at my place to help me. At that time, A helped me call the police, B helped me complete the order, and C helped me negotiate with the driver. At that moment, I felt the warmth and sense of belonging. Our small group is really good. If they need me at any time, I will help them as soon as possible, just like what they did for me.

If we compare individual tactics with collective ones, we can see that the former is mostly aimed at optimizing one's own performance, even to the detriment of colleagues. In this case, couriers behave like many platform users who try to "maximize the benefits of using the algorithmic platforms, including the leveraging of loopholes without directly violating platform rules."[59] These tactics remain mostly aligned with the moral economy of the platform, based on competition, while the collective ones develop mutual aid practices that challenge the competitive logic and operating rules of the platforms.

PRIVATE ONLINE GROUPS AFFORD LEARNING, RESISTANCE, AND SOLIDARITY

Food delivery apps are designed to enable competitive behavior through gamification strategies and a rhetoric discourse based on the neoliberal logic of meritocracy.[60] These apps facilitate direct communication between the worker and the company via their corporate chats, but any functionality that could favor the construction of bonds between couriers is intentionally avoided and any form of peer-to-peer communication is disabled. Through the creation of online private chat groups, however, workers restored forms of mutualism not afforded by the apps. WhatsApp, Facebook, Telegram, and WeChat private groups emerged as indispensable safety nets for online food delivery couriers. They represent informal bazaars where all kinds of exchanges take place that are vital to the survival of every worker. They find in these groups "the comradery and support of their digital colleagues."[61]

Unpacking and gaming algorithms, exchanging tricks, and exploiting platform loopholes—all these activities are enacted through participation on private online chat groups. These private networks represent only one tactic among many, but they are also a device for something else: forging a shared communal struggle. The constitution of a collective consciousness of one's condition of subalternity passes through the thousands of daily conversations that circulate within these private groups. Only by recognizing that the exploitation suffered by one person is similar to that of other peers is it possible to begin to feel part of a community and act collectively. We propose to understand these private chat groups as infrastructures of *learning* (learning environments), *resistance* ("hidden transcripts" of resistance), and *solidarity* (mutual aid and solidarity-building spaces).

LEARNING ENVIRONMENTS

Chat groups are a kind of informal school of food delivery, where the profession of courier is learned. While food delivery companies need a reserve of nonskilled workers—easily replaceable couriers—the information that couriers exchange in chat groups encourages the professionalization of couriers and enables them to resist longer in the market (because they know the tricks of the trade). They can learn and exchange work experiences, share instant traffic information, and any other data that help them survive in this environment. As highlighted by Jamie Woodcock, "The refusal of platforms to provide effective training or support . . . means that workers must resolve many issues themselves. In response, workers seek each other out to share information and discuss the work."[62] In China, for example, newcomers learn from chats not only how to apply for health certificates required by the platforms, but also how to create a fake certificate and save money.

Many of these conversations revolve around working tools and tips on how to make them more efficient. There are a lot of advertisements of equipment for sale (delivery boxes, motorbikes, bikes, e-bikes, helmets, etc), advice on how to repair a moped, or how to save money on gas. Other highly valuable information is about the best areas of cities for people to work and real-time traffic news. For example, when couriers encounter congestion on a certain section of road, they will notify all the others so they can detour around it to avoid wasting time.

Another highly traded piece of information is about the prices of orders. Prices vary all the time and depend on black-boxed parameters. Nobody knows the price received for a same order by other couriers. This information is available only if someone circulates it via the private online chats created by the couriers. In fact, couriers share screenshots of prices in the chats every day to compare the prices for orders in one city to those of another or to see if there are any differences between platforms or various time slots. Through this continuous comparison, couriers learn which prices are above or below the average. This form of collective intelligence (or general intellect, in Marxian terms) allows them to create benchmarks for themselves and set thresholds below which they decide not to accept orders. The discussion of delivery fees is central to every chat and this type of discussion is not limited to food delivery workers; rather, it seems to be a constant in all discussions among gig workers.[63]

Yet the most interesting conversations are those around the workings of order dispatch and personal ranking algorithms. In Turin, Italy, a group of couriers organized by an Italian lawyer meet periodically to understand how food delivery algorithms work. The group of couriers carries out experiments "in the field" with an almost scientific approach:

in each experiment, only one variable is changed (the distance of the courier from the restaurant, the score of the courier carrying out the experiment, or the time slot of the work shift). The aim of this experiment of reverse engineering is to understand what happens in the order allocation system when a single parameter is modified.

On October 11, 2020, a courier from Milan asked in a chat room: "Million-dollar question: has anyone ever been able to figure out exactly why the Deliveroo efficiency value sometimes drops even when you always check in on time, stay within the area and never reassign an order? I'm going crazy trying to figure out what the criteria is behind it!"

This conversation triggered a collective unpacking session of the Deliveroo algorithm, in which everyone explains their theory of how Frank works. These theories, even when they are far from the truth, play a central role in the everyday lives of the couriers because once spread, they influence their behavior and subsequent decisions.

From these conversations, the couriers' "algorithmic imaginary"[64] takes shape (i.e., the set of collective beliefs about how the platforms' algorithms work). These collective beliefs are based on the personal experience of thousands of couriers. Given the black-boxed nature of these algorithms, this imaginary can only be imperfect and contradictory: couriers cannot infer with certainty the operating principles of Frank and his siblings but can only postulate rough theories. Learning the operating principles of food delivery app algorithms, or even just imagining how they work, is a constantly evolving process, shaped by conflicting "folk theories."[65] By discussing it together, couriers increase their "algorithmic awareness,"[66] improving the digital skills that they need to be better workers. This algorithmic awareness, however, is far from homogeneous. The levels of this awareness are markedly different, oscillating from rather accurate (general knowledge of how it works) to vague (automatic execution of platform activities without really understanding them); in any case, intuition always plays a highly relevant part. In Mexico, none of our interviewees were familiar with the notion of an algorithm. In India, we also discovered that even the term "algorithm" was alien to the respondees; nevertheless, all of them were still able to cheat algorithms to some extent by finding loopholes in the platform's services. Also in India, we found a wide digital divide between couriers born in big cities and those working in provincial towns: factors such as education, age, and city and country of residence greatly influence the level of algorithmic awareness of the couriers.

HIDDEN TRANSCRIPTS OF RESISTANCE

These conversations about how to fool the algorithm take place every day, away from the prying eyes of Frank, the algorithmic boss. For this reason, private online chat groups can also be understood as "hidden transcripts" of resistance.[67] In his analysis of everyday resistance practices, Scott[68] traces a subtle difference between forms of resistance that openly manifest themselves in the face of power and those that only do that far from the gaze of power. Where subaltern subjects know that they cannot defeat or subvert power, they maintain a public behavior that is compliant with it, venting their dissent only in places invisible to power. Scott called the first form of resistance "public transcript"—that is, "open action in front of the other side in the power relationship"[69]—while naming the latter a "hidden transcript"—the discourse (both verbal and nonverbal actions) that takes place "offstage" so that powerholders cannot see.[70]

We found the first traces of these hidden transcripts in the conversations that happen among couriers in WeChat private chat rooms. Fearful of being punished by the company administrators who manage the corporate WeChat groups, Chinese couriers prefer to set up or join private WeChat groups in which all members are couriers. While in corporate chats, Chinese couriers never complain, in private chats, they openly express their dissent, aware that they are in a safe environment, far from the panoptic control of their bosses. In their public profiles, gig workers show themselves compliant and smiling, trying to get positive reviews from customers and the platform. In their private chat rooms, however, they pour out their discontent and curse their algorithmic boss.

This online environment is an incubator of future resilience and resistance practices, as Woodcock also argued. According to him, these existing networks can be understood as "the building blocks from which more formal organizations can be developed."[71]

For example, in India, on August 9, 2020, the food delivery platform Swiggy issued an internal communiqué that announced a pay cut for its delivery couriers across at least four cities—Delhi, Chennai, Hyderabad, and Kolkata.[72] Faced with yet another cutback, a small group of couriers created a WhatsApp group. Each of them then invited other couriers to join the group, and together they organized a strike. On August 19 and 20, 2020, more than 500 delivery workers assembled outside the Swiggy office in Malviya Nagar, in South Delhi to protest the pay cut. After this first collective action, the group set up the All-India Gig Workers' Union (AIGWU). On September 15, 2020, Swiggy workers organized through AIGWU went on strike in a number of Indian cities including Hyderabad, Chennai, and Delhi to call for greater pay. They rallied outside of restaurants and prevented third-party companies from picking up orders.[73]

These mass protests might at first appear to be desperate and unorganized actions, the result of sudden outbursts of anger, but they are meticulously organized through the creation of ad hoc private online chat groups.

MUTUAL AID AND SOLIDARITY BUILDING SPACES

Within these groups, each courier builds new bonds and expands their social capital. When a person starts riding for the first time, they often feels lonesome. Adriano, a courier from Messina, Sicily, describes his first few days of work as follows: "This job isolates you a lot because it's you, the bike and the backpack. Then little by little you meet some people and you start to make some friends." While waiting in front of restaurants, couriers start exchanging words, asking each other their names and what areas they usually ride in. Then they exchange their phone numbers. The courier met on the street invites him to join an online chat group: this is the beginning of a potential friendship. The interaction between the street and the online environment is constant: chats are embedded within street life, just as street life is embedded within chats. Chats are an extension of street life, and vice versa. Private chat groups can be understood not only as learning environments and hidden transcripts of resistance, but also as mutual aid and solidarity building spaces. These online environments provide "day to day mutual support."[74] From these online encounters, relationships are born that continue offline and then pollinate into smaller online groups, which in turn provide stronger and more lasting solidarity.

An example of these bonds of solidarity can be illustrated with what happened one evening in December 2020 in Naples, Italy. A courier wrote on a private WhatsApp chat counting about 200 members that a colleague of theirs had been beaten up and his moped stolen. Within two hours, the members of the group opened an online crowdfunding site and raised almost 2,000 euros to allow the robbed courier to buy a replacement moped for himself. Another winter evening, also in Naples, a courier wrote that he had a problem with his moped: it had stopped and would not start up again. Two other couriers responded immediately and interrupted their work to pick him up and take him home. Antonio, a courier from Naples who had joined the left-wing Italian national union, the Confederazione Generale Italiana del Lavoro (CGIL; or Italian General Confederation of Labor), told us that "in Naples, no one is left behind" and worked to establish Casa del Courier (Couriers' Home), a communal public space where couriers could meet, take a shower, rest, and chat with compadres.

In Mexico, a group of ten drivers has built a series of mutual support tools to exchange information and help each other in case of emergency. The group operates with two group chats on WhatsApp: one is aimed at exchanging internal information, notifications on street traffic, entertaining content (memes, jokes, videos, and gossip), while the second is strictly focused on the security and safety of the members of the group: everybody reports when they start or end work; as soon as they start, they send their UTR (the Spanish acronym for *Ubicación en Tiempo Real*, or, "real-time location") to the group, lasting eight hours (the maximum allowed); emergency information (worker's name, car data, emergency contact, etc.) and security warnings (accidents, threats, assaults, etc.). In addition, they use the Zello app[75] to communicate using radio codes (mostly numbers). The drivers keep each other company when morale is low.

The practice of creating self-defense and support groups for emergencies is also popular in South Africa, where drivers who use Bolt, an Estonian Uber-like app operating in 500 cities, have been digitally crowdsourcing their safety nets on WhatsApp and Telegram. In these groups, they alert other members when they are in danger or share their location in real time. Bolt drivers have suffered several deadly robberies and said that the company's safety solutions "just don't cut it,"[76] so this feature offers some measure of security.

ENTREPRENEURIAL VERSUS OPPOSITIONAL ALGORITHMIC SOLIDARITY

These private online environments enable the building of bonds of solidarity among workers. This solidarity, however, is so extensively shaped by algorithms that we might describe it as an *algorithmic solidarity*. By this term, we mean all those forms of worker solidarity that are mediated *by* algorithms or that emerge *around* algorithms. An example of algorithm-mediated solidarity involves those food delivery couriers that discover new tricks of the trade through videos of other more experienced couriers that are suggested to them by the algorithms of Douyin (known as Tik-Tok outside of China) or YouTube. These videos allow couriers to connect for the first time with other workers like them. In China, for example, the members of the Knights League (KL)[77] WeChat group, a spontaneous courier union founded in 2018, also recruit new members via Douyin. Searching for courier groups online is also mediated by algorithms. Workers sometimes receive an invitation from one of their colleagues to join one of these groups, but other times, they discover these groups through algorithmic suggestions.

Solidarity *around* algorithms, on the other hand, indicates all those forms of cooperation and mutual support that are emerging to face the power of the algorithms. This includes the sharing of resources and information to better cope with the negative implications of platform power or tricks and recommendations to use algorithms in ways that benefit workers and can generate more effective forms of resistance *through* algorithms.

The solidarity that emerges from the intersection of offline (streets and squares) and online (private chat groups) interactions represents a safety net for workers and increases their resilience. But can it really be considered a form of radical resistance to the power of platforms? Or is it not, rather, a way to survive in this precarious work ecosystem with the sole aim of increasing the earnings of all community members? Indeed, in many cases, the algorithmic solidarity that we have observed is nothing more than a network of mutual aid, with no political dimension or intention to radically change their working conditions. This is the case, for example, in solidarity log-out practices, as couriers help each other to reach daily productivity targets. Similarly, in the case of the exchange of working hours or collective unpacking sessions on the workings of algorithms, couriers contribute their little pieces of knowledge to generate a deeper understanding about the mechanisms of their work, while also producing more value for companies. This kind of solidarity, though by no means encouraged by platforms, simply aims at the collective optimization of one's entrepreneurial skills, as has been noted by other scholars in other types of gig work. The Filipino scholars Cheryll Ruth Soriano and Jason Vincent Cabañes, for example, call this "entrepreneurial solidarity."[78] In their study of Filipino freelance platform workers, Soriano and Cabañes highlighted how entrepreneurial solidarity emerges through the Facebook groups created by the workers. They argue that this kind of solidarity not only empowers workers with a sense of agency, but also tends to enhance their entrepreneurial spirit and "serve(s) to dampen possibilities to meaningfully challenge the structures of power underlying digital platform labor."[79] Entrepreneurial solidarity is quite common also among food delivery couriers, as we have already seen. However, in addition to this form of solidarity, we have noticed a critical and oppositional type of solidarity, which is not aligned with the moral economy of food delivery platforms. Examples of this kind of solidarity include collective rejection of orders to protest too-low delivery prices, actions to sabotage deliveries, and building stations where people can meet, repair their vehicles, and build social bonds.

This solidarity among workers is geared not only toward improving their working conditions, but also toward changing these conditions through organizing protests and strikes, or creating alternative delivery platforms, owned by couriers' collectives.

TWO COMPETING MORAL ECONOMIES

The dichotomies of platform work—low pay but occasional freedom, huge insecurity but easy money—show why it endures and continues to grow in spite of the many criticisms of the model. Thousands of gig workers around the world investigated by a global survey revealed mixed feelings about their job: almost 65 percent of gig workers said they often felt happy at work in the past week; however, many frequently felt worried, unsafe, tired, or angry, and 49 percent of them said that they had participated in strikes and protests in the past. Almost everywhere, delivery was perceived as the most stressful type of platform work.[80] We found the same mixed feelings among the couriers we interviewed—so much so that we can speak of the existence of two competing moral economies.[81]

Couriers can be divided into two kinds of "parties": those who adhere to the moral economy of the platforms, espousing their principles and justifying their actions, and those who resist this economy. Of course, this opposition is never clear-cut and stable over time: there are couriers who have changed their positions back and forth throughout their working lives, becoming more or less critical, and there are also those who only partially adhere to the moral economy of the platform. Couriers are positioned along a continuum, at the ends of which are two distinct ideal types of moral economy: that of the platforms and that of the couriers. But what do these two moral economies consist of? The first is based on the neoliberal logic of free competition between self-entrepreneurs and the ideology of meritocracy: those who "work hard" deserve to be given more orders than others. The second is based on a cooperative logic, which places mutual aid at the center and justifies illegal actions if they help to improve collective working conditions. This second moral economy also recognizes that the meritocratic ideology is a red herring because the algorithm is discriminatory and does not allow everyone to play under the same conditions.

COURIERS WHO SIDE WITH THE MORAL ECONOMY OF THE PLATFORM

These couriers adhere to the platform's meritocratic narrative. They are constantly aimed at optimizing their performance and eventually can also practice individual forms of "gaming" (booking hours via bots, multiplication of accounts) to the detriment of other couriers. Most of them are *stakhanovists*[82] who ride at least ten to twelve hours every day, six or seven days per week. In the chats, they brag about how many hours they work and share screenshots of the kilometers they covered or of how many orders they received in a single day.

This group is convinced that if a courier does not receive any orders, the blame lies with him or the area where he decided to hang out. They suggest that maybe such couriers don't know how to work (or don't want to), or there are too many couriers and not enough orders available. They never, ever, question the platform or the algorithm, which they consider to be highly meritocratic and fair: "This is a real meritocratic job, if you work hard, you get money back" (Adrian, August 20, 2020). Stefania, a female courier from Milan, once wrote in a chat: "If you are a real courier, you know that if you work honestly, you bring in the dough. And you don't need to cheat because you're a good courier. And you have high stats. You're well respected by the company and if you need a hand, they'll give it to you. And you don't need to be a smart ass and stick it up your colleagues' butts." If others point out to them the injustice of a ranking that gives many orders to few couriers and few orders to many couriers, they reply that "you have to work hard for your money."

Most of them want to believe in the platform's promises even when earnings fall. This blind faith in the ideology of commercial platforms stems from a proclaimed strong love for "freedom." They place an extremely high value on individual freedom and want this work to remain "autonomous" and unregulated. They frequently have had previous job experience as subordinate employees in factories, restaurants, or logistics and left these jobs for an unstable and precarious one who let them feel "free." There are many who think this way, especially in Italy and Spain: these couriers often lean toward right-wing political ideologies or have a general distrust of political parties and workers' unions. They claim that they feel free while doing this job because they can move within the city playground instead of making repetitive actions in a factory—they don't have a boss who is always on their back. They do this job for the freedom that it apparently grants them,[83] not only for the money they can make from it.

COURIERS WHO REJECT THE MORAL ECONOMY OF THE PLATFORM

On November 30, 2020, an Italian courier, Christian, writes in a chat room: "It's OK to earn money, but it's not OK to over-earn on the backs of the couriers. That means being a piece of shit."

In their private chat groups, couriers talk about how fair or morally acceptable the price of a single delivery is. They also talk about the "lack of respect" that couriers get from platforms and discuss what is the minimum price that a delivery proposal can be and still be decent. Another Italian courier, commenting on the introduction of free login by Deliveroo and the decrease in the hourly wage, wrote on December 30, 2020: "There is something more important than money and it is called dignity."

On December 18, 2020, an Italian courier who was angry with Deliveroo because, according to him, it had not yet paid him the 150 euros it owed him, wrote in a chat: "They fuck me, I fuck them. They play dirty, I play dirty with them."

On December 28, 2020, Deliveroo sent all its couriers a video thanking them for working hard during the first pandemic year. The video was reviewed in a WhatsApp chat room hosting several hundred Italian couriers in different cities and received ironic and unenthusiastic comments. The couriers perceive the sidereal distance between the rhetoric used by Deliveroo—in which couriers are called "heroes"—and the "starvation wages" that the same company is used to paying them. Some couriers compared the cloying rhetoric of Deliveroo's video to "love bombing," a psychological technique used by narcissistic personalities to emotionally manipulate people. Within hours, the video had been *detourned* into a satirical meme.

Those who criticize the platform do so mainly for three reasons: low pay/income volatility, lack of workers' rights (paid holidays, sick days, accident insurance), and black-boxed algorithms. They are usually inclined to consider any collective tactics that may harm the platform as legitimate defensive weapons. Uber drivers think the same way: "Uber screws us over every day and so we have to find creative ways to work around this to make it fairer for drivers."[84]

Yet there is no clear boundary between one moral economy and another, and couriers' positions can swing to one side or the other over time. Most couriers in fact constantly oscillate between these two moral economies. Theirs, we might say, is a "negotiated" position, not entirely adhering to either the hegemonic position of the platforms or the resistance one. Couriers in this position are not opposed to the moral economy of platforms per se, but they question it when it reduces them to "starvation" or to extremely poor working conditions. This attitude reminds us of that of the British rioters studied by Edward P. Thompson. The British scholars Andrew Charlesworth and Adrian Randall, in reconsidering the importance of the concept of the moral economy, remind us that the rioters were not against the free market per se and did not complain so long as this system guaranteed them the ability to buy food at acceptable prices, but "when the 'capitalist market imperative' was permitted to 'drain the country' . . . the larger imperative of the moral economy was invoked and the crowd intervened to safeguard their own supplies."[85] The same is true for many couriers who have joined the protests: the appeal to the immoral behavior of the platforms and the request of a minimum hourly wage occur only when their income drops dramatically.

BEYOND ALGORITHMIC RESISTANCE: EMERGING ALTERNATIVES TO COMMERCIAL PLATFORMS

Nadim Hammami (see figure 3.10) is thirty-three years old and holds a bachelor of arts degree in sustainable tourism management. He started riding for Just Eat and Deliveroo in 2016, while he was studying in Florence, Italy. In 2021, tired of working for these platforms, Nadim joined a group of Florentine couriers to found Robin Food, a local food delivery cooperative. They claim that their initiative is "green, ethical and local."[86] Robin Food is the first Italian co-op to join the international federation of bike delivery cooperatives, Coop Cycle.[87]

What distinguishes Robin Food from commercial food delivery platforms? It claims that it wants to support democracy in the workplace and guarantee the dignity of the worker without neglecting the local economy. Nadim embarked on this initiative only after experiencing the exploitative conditions that the platforms subjected him to. Tactics, tricks, and strategies are effective to survive in the platform ecosystem, but for him, they were not enough. So he tried to create an alternative to them. However, Nadim is not alone. It is amazing how, just a few years after the founding of food delivery start-ups, so many cooperatives have sprung up around the world and developed alternative digital platforms driven by ethical values. In Spain, for example, there are Botxo couriers (Bilbao), Zampate Zaragoza (Zaragoza), Rodant (Valencia), Eraman (Vitoria),[88] La Pajara ciclomensajeria (Madrid), and Mensakas (Barcelona).

3.10 Nadim Hammami, a thirty-three-year-old courier and cofounder of the co-op Robin Food, started in Florence in 2021. Photo by Tiziano Bonini.

Most of them are members of Coop Cycle, too. For them, it is essential that labor rights and adequate working conditions are guaranteed, as well as paying their taxes in the countries where they operate.

The founders of these co-ops all share a previous background as couriers working for commercial platforms. They learned all the individual and collective tactics that we have described so far, created bonds of solidarity with other peers, and developed a greater awareness of their own condition of subordination and finally decided to change it. Of course, it is not easy for these co-ops to become sustainable. The model of these platforms differs from that of commercial ones in their desire to generate an economy of proximity and guarantee decent working conditions. This means that the service provided by these co-ops has a slightly higher cost for the end customer, but as a Spanish courier, Pepe, pointed out: "The cheap price for the customer is where the precariousness lies."

Does platform cooperativism[89] represent a real alternative to commercial food delivery platforms and a more concrete form of resistance to their overpowering?

It is still early to say. We cannot know whether Coop Cycle, Robin Food, or Mensakas are models capable of sustainable growth while continuing to guarantee secure employment and more democratic participation in the management of companies. However, according to Trebor Scholz, the founding director of the Platform Cooperativism Consortium at the New School of New York, "Platform co-ops offer a more democratic and equitable alternative to traditional companies, and they have the potential to create good jobs, boost local economies, and increase resilience in the face of future shocks."[90] In a postpandemic world, Scholz argues that "platform cooperatives could help to build a fairer, more sustainable economy that works for everyone."[91] Scholz estimates that there are around 550 projects in forty-three countries, spanning industries like short-term rental, transportation, domestic work, health care, and energy, but the real figure could be much higher.[92]

Many of these attempts will not last, but their mere existence shows that gig workers are by no means passive in the face of the computational power wielded by platform capitalism. Everyday resistance tactics, tricks and stratagems, and riots and strikes are only the beginning of a new wave of resistance that is becoming more and more structural.

4

GAMING CULTURE

INTRODUCTION

A release of the Korean pop music sensation BTS has pushed its fans, millions of people around the world, to help them circulate its album, launching a sophisticated campaign (or "stream party") to make sure that the boy band reaches number 1 on the charts. BTS fans in the US created fake accounts to play the band's music on music streaming services and distributed access to users' accounts in other countries via social media. Recipients streamed BTS music continuously over multiple devices through virtual private networks (VPNs), which can "spoof" locations by redirecting user traffic through multiple global servers. Some fans would even go as far as to organize donations so that other fans could pay for premium streaming accounts. This type of networked action artificially inflated the consumption of the album and tricked Spotify's algorithms, which interpreted the rapid increase in clicks as a fast-rising music trend deserving greater visibility. In turn, Spotify's algorithms automatically included BTS songs in a greater number of algorithm-generated playlists. In this way, Spotify's algorithms acted as unwitting allies of the strategy designed by BTS fans.

This is just one example of the practices implemented by consumers of cultural artifacts in their attempts to manipulate algorithms for their own benefit. In chapter 3, we saw how gig workers—and, more specifically, food delivery couriers—are able to exercise some limited, but also very creative, forms of agency vis-à-vis the power of the platforms they work for. In this chapter, we will focus on how algorithmic agency is exercised in the field of platformed cultural industries, by both producers and consumers of cultural objects. We will explore how algorithms are resisted in the realm of cultural production and consumption by examining the case of independent content creators of the global influencer marketing industry. More specifically, we will investigate Instagram "engagement groups," where users exchange "likes" to artificially manipulate their visibility on the platform. First, we have to understand how the rise of platforms is changing traditional cultural industries, while reshaping cultural creation, distribution, and consumption.

THE PLATFORMIZATION OF CULTURAL INDUSTRIES

The creation and circulation of cultural artifacts have long been mediated by professionals in the various cultural industries. However, the rise of digital platforms is transforming the role of cultural industries and turning culture into an increasingly platform-dependent commodity. Two media scholars, David Nieborg and Thomas Poell, call this process "platformization of cultural production," which they define as "the penetration of digital platforms' economic, infrastructural, and governmental extensions into the cultural industries, as well as the organization of cultural practices of labor, creativity, and democracy."[1] In another work written with José van Dijck, the two authors conceive of this process as the "reorganization of cultural practices and imaginations around platforms."[2]

But what does this reorganization entail? It refers to the ways in which content producers and consumers modify their activities, imaginations, and identities to comply with the logic of global digital platforms.

According to media scholars such as Brooke Erin Duffy and her colleagues, we should understand this process both as institutional and as "rooted in everyday cultural practices."[3] Poell, Nieborg, and Duffy[4] identify three dimensions that characterize this process of platformization of cultural industries: markets, governance and infrastructure.

SHIFTING MARKETS

Platformization brings about a shift from single and two-sided markets to complex, multisided markets. For example, Spotify and YouTube act as intermediaries among different actors (music and content creators who upload their work on these platforms, advertisers, listeners and viewers, public and private institutions, cultural intermediaries, entertainment companies, etc. . . .), or sides of the market in which they operate. According to Poell and his colleagues, platforms act as "matchmakers by connecting consumers or 'end-users,' a wide variety of businesses (advertisers, content creators, etc.), governments, and nonprofits."[5] The platformization of the cultural industries—from music to television, from publishing to journalism, from fashion to design—strongly affects the positions of power acquired by traditional cultural producers. For example, newspapers, which so far have played a central role in the intermediation of information, are being marginalized since they increasingly depend on the algorithms of Facebook and Twitter to deliver the news that they produce for their readers. This change triggers the adoption of new managerial strategies by legacy media and pushes them to adapt their publishing mechanisms to the logic of the platforms.[6] For example, public service radio stations have to decide whether to make their podcasts available on Spotify and reach a larger audience or keep them on their own website.

CHANGING GOVERNANCE

Platformization not only affects the markets of cultural production, but also the ways in which cultural production is governed. Platforms decide the ways in which the audience discovers cultural artifacts and the rules that govern the processes of cultural distribution. Photographs, newspaper articles, personal comments, and videos are all subject to the rules of the platforms, which have the power to accept them without reservation, reject them (by banning them), or accept them but limit their visibility. The story of Facebook's censorship of the famous photo of the naked Vietnamese girl running down a village street, evoking the horrors of the Vietnam War, is a classic example of this normative power.[7] Nieborg and Poell argue that platforms exercise "significant political economic and infrastructural control over relations between complementors and end-users."[8] Poell et al. in particular focus on forms of "governance *by* platforms" that "structure how content can be created, distributed, marketed, and monetized online, affecting the regulation of public space more generally."[9] These authors show that the governance choices of platforms have an impact on the autonomy of cultural producers.

INFRASTRUCTURAL TRANSFORMATION

Platformization transforms the *infrastructure* of cultural production. This means that creators who have shaped cultural artifacts through their own creativity must necessarily make the artifacts suitable for the platform that will provide the infrastructure to connect them with a potential audience. Cultural producers will therefore have to *optimize* their creations, preparing them so they can be as "recognizable" by the platforms' algorithms as possible.[10] The American media scholar Jeremy Wade Morris underlined the increasing pressures on artists and producers to make a "Spotify song,"[11] defined as "music that seems sonically optimized for Spotify's platform and for the various listening occasions and environments for which users turn to Spotify for sonic accompaniment."[12]

Nieborg and Poell argue that cultural producers "are incentivized to change a predominantly linear production process into one in which content is contingent, modularized, constantly altered, and optimized for platform monetization."[13] According to Nieborg and Poell, not only are cultural industries dependent on platform power, but also platformized cultural artifacts are *contingent*—that is, they are continuously reprogrammed and manipulated, as the position of a song in a playlist or the headline of a news item on an online newspaper.

When the infrastructure provided by platforms becomes the main gatekeeper of their visibility, then content creators have to adapt their creative ideas to the technological affordances of the platforms. This phenomenon is not entirely new, since content creators always had to adapt to previous media infrastructures, such as radio and television.

In sum, the idea of the platformization of culture argues that cultural producers, cultural artifacts, and cultural consumers are increasingly dependent on the choices of platforms that become the new gatekeepers of the cultural industry.[14] These new gatekeepers react in real time to the consumption choices of their audience and modify the artifacts to meet their needs and wants, or even better, to anticipate them through an accurate analysis of their taste profile.[15]

If the form and relevance of cultural artifacts are contingent, and not immutable, it also means that they can be manipulated. Usually, this manipulation is the prerogative of platforms, which modify their products according to user feedback. For instance, Netflix offers the same film or television series to different users, adapting the cover according to the user's taste profile: in this way, the same object can take manifold forms based on distinct cultures. Yet if a cultural object is contingent, its visibility also depends on user feedback. This then means that even the user, to a lesser extent, has the power to (organically or artificially) affect the visibility of that artifact, as in the example of the K-pop fans given at the beginning of this chapter. This is what this chapter will focus on, after clarifying how cultural work is changing with the rise of platforms.

PLATFORMIZED CULTURAL WORK: PRECARIOUSLY SEEKING . . . VISIBILITY

What does it mean to work in the cultural and creative industries today? What is cultural work in the time of platform capitalism? In recent years, many scholars have focused their research on the changing nature of cultural work, describing it as "fraught with stress and burnout,"[16] increasingly precarious and insecure, less unionized, and more individualized.[17] In some ways, however, working in the cultural industries has always been characterized by a certain amount of precariousness and insecurity. Those who have read the novel Lost Illusions by Honoré de Balzac[18] may remember the story of the provincial young man Lucien Chardon. Chardon arrives in Paris under the illusion that he could become a famous poet. He desperately seeks fame, but all he gets are disappointment and a precarious artistic life. Balzac describes with ethnographic precision the bohemian life of the Parisian creative class of the 1820s (e.g., writers, journalists, poets, painters, printers, theater critics, actors, and publishers). It is like reading the chronicles of today's creative workers: penniless, forced to go out every night and hustle to meet people who could give them a job, one month rich, the next month broke.

While it is true that careers in the creative industries have long been marked by precarious conditions of work and unstable wages,[19] the economic model imposed by digital platforms are challenging the survival model of cultural and creative workers. Brooke Erin Duffy and her colleagues argue that today's platformized creative economy amplifies the precarity of the past because it is powered by a neoliberal ethos that values self-commodification and recasts independent employment as self-entrepreneurship.[20] What distinguishes platform cultural work from previous forms of cultural work is an even more extreme and individualized condition of precariousness and, above all, a more intense "metrification" of the worker's performance: fame, popularity, and visibility are constantly monitored, calculated, and updated. Just as visibility can fluctuate due to a change of the metrics that measure it, the income of a platformized creative worker is fragile, ephemeral, and volatile. Careers in this field are "bound up with fluctuations in wider socio-economic, cultural, and political realms,"[21] like the price of gold on the stock market. Visibility has become the main currency of platformized cultural and creative works. To make a living, a freelance cultural worker—ranging from a musician to a journalist publishing on Spotify or Medium—has to accrue more and more visibility. Cultural labor in the age of platforms is a "visibility labour"[22] or a "visibility game,"[23] but cultural workers do not have full power over their visibility; it is not a commodity they control. On the contrary, it depends on factors external to them.

To a certain extent, fluctuations in the visibility of cultural work have always existed, but what is new today is the ability of platforms to capture and govern the visibility of cultural workers. By developing the technical infrastructure that affords them to quantify the visibility of a cultural product, platforms have been able to subsume the visibility of cultural work, to fence it under their boundaries and turn it into a new, precious commodity. Visibility has been appropriated and datafied: the computational infrastructures built by platforms continuously transform creators, cultural objects, and consumers into data. The visibility of a song, for example, is quantified and broken down into dozens of data points, such as the number of plays within a specified period of time, the number of people who have saved the track in their library, the average time spent listening to the track, and the "skip rate" (i.e., the number of people who have stopped listening to the track before the end).[24] The same process of metrification is applied to every digital cultural artifact, like online news articles, films, podcasts, and ebooks. Creators and consumers are data-fied too. Every subject that takes part in the platform ecosystem is being metrified to establish her position in the visibility rankings. As cultural work is increasingly turned into a data-driven activity, cultural workers are pushed to "pursue quantifiable markers of visibility."[25] Duffy argues then that visibility has become a requirement for career success in cultural industries amid platformization, and accordingly, "metrics are the central axes on which power and resources are exchanged—in the form of sponsorships, brand deals, collaborations."[26]

Platforms possess the means of production (data and algorithms) to capture and govern the visibility of their complementors. While in the case of the gig workers discussed in the previous chapter, the computational power of platforms served to govern the performance of gig workers' *material bodies*, in the field of cultural industries, the computational power of platforms serves to govern the performance of cultural workers' *immaterial creativity*, expressed through the currency of visibility. Visibility is thus the battleground where platforms and cultural workers confront each other. In the platform ecosystem, algorithms exert disciplinary power over users through what Taina Bucher has called "the threat of invisibility":[27] unlike the Foucauldian Panopticon, where prisoners feared being constantly visible, Bucher argues that what platform users fear is losing their visibility. This threat pushes users to implement practices, legitimate or not, aimed at being algorithmically recognizable: Instagram users know that if they want a photo to be detected by the algorithm as a potentially viral image, they need to get a certain number of "likes" and "shares" within the very first few minutes after posting it. If visibility is so central to the survival of cultural workers offering their products on platforms, it is understandable that they will do anything to gain more of it, including practices aimed at gaming the platform's algorithms. Like the gig workers from the previous chapter, they try hard to understand the functioning of the algorithms that govern them, generate theories to explain their behavior, engage in reverse engineering practices, and develop tactics to cheat them.[28] However, this process of algorithm sense-making belongs not only to cultural workers, but also to ordinary cultural consumers such as music listeners, Netflix viewers, and Instagram users, and to commercial agencies like an emerging cottage industry selling tricks and "hacks" to game an algorithm.[29]

Platform power in the cultural industries, although increasingly endemic, is neither monolithic nor stable over time. Indeed, as Nieborg and his colleagues argue, we should develop a nuanced and ambivalent understanding of this power: "Platforms exert mechanisms of power over the phases of the creation, distribution, monetization, and marketing of culture; but they also furnish space for negotiation and contestation."[30] The prosumers of platformized cultural industries exert strategic and tactical algorithmic agency on their online visibility. Wherever visibility is at stake, we find individual and collective practices that attempt to artificially manipulate this visibility. Again and again, where there is power, there is resistance. Negotiation and contestation of algorithmic computation can, in fact, be detected in all the fields of platformized cultural industries, from music streaming platforms to Instagram and TikTok.

GAMING PRACTICES ARE EVERYWHERE

During the first COVID-19 lockdown in Wuhan, China, in March 2020, millions of Chinese students had to stay at home and attend classes online. One of the most popular platforms used by schools in Wuhan during the lockdown was DingTalk, a Chinese app for collaborative online work created in 2016 by the Alibaba group to "make work and study easy." Students had to download the app and log in to attend classes, and teachers used it to send homework to the students. At one point, however, the app risked disappearing from the Chinese App Store, as its score dropped from 4.9 to 1.4 overnight.[31] Somehow, hundreds (maybe even thousands) of Chinese students had orchestrated a coordinated effort to review the app negatively and consequently damage its ranking. The Chinese students realized that by flooding the app with negative reviews, its reputation would drop so low that it would risk being kicked out of the App Store or lose visibility in its search engine.

Attempts to cheat the algorithms of platforms in the cultural and creative industries are everywhere; that is, they are an *endemic* feature of the platform society. The discoverability and visibility of platformized cultural items are qualities that are never completely determined by platforms, but they are the result of a continuous and dynamic contest. Visibility in the platformized cultural industries is a contested commodity. On one side of the barricade, there is the power exerted by platforms; on the other side, there are gaming tactics that can be found in any platform environment, from online video games[32] to fitness apps,[33] from online dating apps[34] to podcasting[35] and music.[36] Music industry reporters, in particular, have shed light on many stories about music producers who were able to artificially inflate their ratings and delude Spotify's algorithms into thinking that they were dealing with genuinely emerging hits that deserved a place on the more popular playlists. The motivation for gaming Spotify's algorithms in these cases was purely economic: they mobilized both computational power and substantial budgets to generate an impressive increase in the number of "plays" and be rewarded with more visibility and more revenues. Spotify is highly critical of these kinds of practices, which it interprets as "artificial streams," and it has taken several countermeasures and sanctions against those users who are doing this,[37] but it has not yet succeeded in curbing this practice.

Yet attempts to cheat algorithms don't just come from music industry professionals. They can also be carried out "from below" (i.e., by communities of fans and listeners who are not motivated by a desire to increase their income, but by passion for their idols). As we saw in the opening story of the chapter, this is the case with K-pop fans. These fans are famous for their incredible ability to mobilize, including planting a rain forest for their favorite artist, buying expensive ads in Times Square, or sinking US president Donald Trump's rallies.[38] In 2018, the global fandom of the Korean band BTS became famous for being able to manipulate Spotify's streaming charts. Upon the release of its album *Love Yourself: Answer* in August 2018, superfans of the band launched a sophisticated campaign to make sure that it reached the top of the Billboard charts. BTS fans who were in the US apparently created accounts on streaming platforms and distributed the login details to BTS fans in other countries. The recipients would then stream BTS's music continuously by VPN, often with multiple devices. By using this strategy, one BTS fan group claimed that it had distributed more than 1,000 Spotify accounts, boosting BTS's ranking on the

US Spotify music charts.[39] This tactic is so widespread that there is even a Korean fan-slang word for streaming an artist repeatedly to boost chart numbers: *sumseuming*, or "streaming 24/7 as one breathes."[40]

The cultural studies scholars Qian Zhang and Keith Negus argue that these practices are the result of a particular kind of fandom: the "data fandom."[41] Data fans, according to them, adopt individual and collective strategies to "deliberately intervene and to influence the statistical, sonic and semantic data collected by and reported on digital platforms and social media."[42] These fans have realized that they are relevant to the music industry in that they can be turned into data, so they use this knowledge to benefit their favorite musicians and enhance their sense of achievement and agency.

We understand these practices of gaming algorithms "from below" as manifestations of algorithmic *tactical* agency that are not aligned with the moral economy of streaming platforms. In fact, what is considered illegal by Spotify and other streaming platforms is instead deemed as morally acceptable to fans: "We're just trying to support our idols," says Lamia Putri, a BTS fan from Yogyakarta, Indonesia. On the contrary, fans consider *Sajaegi*—a practice orchestrated by Korean marketing agencies for profit—to be unfair. *Sajaegi* means "unethically and/or illegally boosting a chart ranking."[43] K-pop fans consider *sumseuming* a fair practice, whereas *Sajaegi* is wrong because "it's outside our control. It's created by people who have power. We're only fans. We can't do anything but get angry."[44] Yet *Sajaegi* is not a native practice of the platform era, since it existed even before when entertainment agencies used to bulk-buy their own artists' CDs. In the digital era, they are using bots or troll farms to repeatedly stream songs and raise chart numbers. Many Korean bands today receive offers of *Sajaegi* from marketing agencies to inflate their audience in exchange for money. According to the moral economy of streaming platforms, both *sumseuming* and *Sajaegi* are illegal, while the moral economy of K-pop fans distinguishes *sumseuming* from *Sajaegi* and considers the latter just a greedy manifestation of the power of money.

While the practice of *Sajaegi* is an example of strategic algorithmic agency that is not aligned with Spotify's moral economy, *sumseuming* is an example of tactical algorithmic agency, and it also fails to comply with the platform's moral economy. Yet this agency can be detected in all spheres of the platformized cultural industries, particularly on those platforms such as YouTube, TikTok, and Instagram, where the income of independent content creators depends entirely on the "vagaries" of the algorithms that determine their visibility. In these cases, we have noticed that independent content creators have developed very sophisticated collective tactics to game the algorithms. Social media content creators are an important case study for understanding the lived experiences of platformized creative workers. As Zoe Glatt explains, "whilst most cultural industries existed prior to the Internet, the influencer industry is a rare example of one that grew out of it."[45] Therefore, we decided to focus on a specific kind of platformized creative worker— independent content creators on Instagram—and study[46] their tactics to survive amid the rising uncertainty of the global influencer marketing industry.

GAMING INSTAGRAM: AMONG INSTAGRAM ENGAGEMENT GROUPS

"Done. Put a like in my profile picture, please!" "Done. You can reciprocate my like by putting a like on the picture you prefer." "Done. Reciprocate it? Will you help me liking my wife and daughter's picture? You'll find it among the first comments in this post, thank you all!!!"

These are just some of the comments (see figure 4.1) you might come across if you join an Instagram pod, also called an *engagement Pod*, an *engagement group*, or a *direct message (DM) group*.

According to the new media scholar Victoria O'Meara, who was among the first to research this phenomenon, engagement pods are "grassroots communities that agree to mutually like, comment on, share, or otherwise engage with each other's posts, no matter the content."[47] Facebook, Instagram, Telegram, and WhatsApp are the most used platforms to host these groups. The term "pod" means both "a herd of marine mammals" (including whales, dolphins, walruses, and seals) and "a vegetable casing." The idea of a pod already suggests a group of individuals acting collectively in the same way, driven by common interests. The members

4.1 An Italian Instagram engagement group, or pod. Red circles highlight the requests made by the members to their peers: "Reciprocate," "Put a like to my profile photo," "Reciprocate, please," or "Follow me."

of an Instagram pod are all united by the same goal: to increase their visibility on the platform and avoid paying for sponsored content.

Pods become popular to the wider audience only in March 2016, when Instagram replaced its chronological feed with an algorithmic one. On March 15, 2016, the popular social media platform announced the implementation of algorithmic personalization on its official website as follows:

You may be surprised to learn that people miss on average 70 percent of their feeds. As Instagram has grown, it's become harder to keep up with all the photos and videos people share. This means you often don't see the posts you might care about the most. To improve your experience, your feed will soon be ordered to show the moments we believe you will care about the most. . . . [48]

Soon after this change, content creators across the Instagram community began reporting diminished reach and engagement numbers. Many users were angry about this change and created the hashtag #RipInstagram.[49] A Change.org petition demanding that Instagram revert its news feedback to chronological order garnered 343,011 signatures.[50]

O'Meara interpreted engagement pods as a "grassroots tactical response developed to contend with these algorithmic conditions of cultural production."[51] Instagram considers these groups illegitimate and works constantly to track and eliminate them. Point 17 of Instagram's Platform Policy explicitly advises users not to "participate in any "like," "share," "comment," or "follower" "exchange programs," while in the platform's Community Guidelines, there is a call to not "artificially collect likes, followers, or shares." Yet these calls are useless because since 2016, engagement groups have multiplied and became endemic to the platform. Engagement groups, or "pods," are common among TikTok users and creators too, but their utility is contested.[52]

HOW PODS WORK

Pods sprout like mushrooms on several platforms, not just Instagram. Their life is as fragile as a soap bubble or a butterfly. They grow fast, and just as quickly they burst: they have an ephemeral and contingent life, like the cells of a secret association. Their maximum size depends on the rules established by the platforms. Closed groups on Facebook have no membership limit, while WhatsApp groups cannot exceed 256 members. On Telegram, groups can reach a maximum of 200,000 members, which is why pod administrators usually employ a bot to patrol its many members. Each group is administered by one or more people committed to set the rules of the group, verify that they are respected, and accept or reject new members. Pods vary greatly from platform to platform and may be created for various purposes: to exchange follows, "likes," or comments, or all of them altogether.

While some pods are organized around a common theme, others are dealing with every kind of content. Ruben, a twenty-nine-year-old microinfluencer living in Milan, stated that in the beginning, he took part "in groups where there was really everything," an expression that he used to differentiate them from what he defined "topic-oriented groups";

Abramo, a microinfluencer from Verona, also spoke about "generalist" groups, as opposed to "niche" or thematic groups. In the second case, they can focus on a common passion, like photography, fashion, or food. This difference is very important, as we will see, because in these topic-oriented, usually niche groups, the degree of support, mutualism, and solidarity is higher.

An extremely specific type of pod is the *round pod,* found only on Telegram, which is organized by turns (known as "rounds") and administered through bots. The bot that manages the round pod randomly selects a minimum number of members that at a specific hour must *drop* (i.e., share their Telegram profile with the group). Half an hour after the drop, the round starts: all the members of the pod have to put a "like" on the last photo published by the users selected by the bot. They only have half an hour to comply with the task, or else the group will ban them. Round pods last for only a limited time. Each round has a specific duration because its members developed specific theories regarding how the Instagram algorithm calculates the engagement rates. This sort of reverse engineering of the temporality of the algorithmic detection process of engagement is based on a complex mix of word of mouth, collective observations, and discussions within the pods. Altogether, these conversations form a "theory" coming "from below," a kind of collectively assembled knowledge that Motahhare Eslami and her colleagues call "folk theories."[53] It doesn't matter if this belief is true or not: so long as these theories are *perceived* as true by the group, they will shape the behavior of the pod members. The algorithmic agency of the pod members is strongly shaped by these folk theories.

Pods are populated both by common users (i.e., those who have a small number of followers or who have just landed on the platform) and various types of experienced influencers. It will be very difficult, however, to find macroinfluencers and megainfluencers with hundreds of thousands, or even millions, of followers among the members of these groups: macroinfluencers—established stars in the platformized cultural and creative industries—do not need to rely on these groups for visibility. Yet among the members of a pod, we can find many of those influencers that Crystal Abidin[54] and Victoria O'Meara[55] call *nanoinfluencers* and *microinfluencers*[56] (i.e., people who have an audience of a few thousand followers and have just started their careers as professional content creators or are not yet doing this job full time). These types of cultural content creators have a constant need to increase their visibility because wider audiences allow them to attract better sponsorships.

Nanoinfluencers and microinfluencers often resort to pod support to achieve the engagement goals set by their sponsors. Without pods, many influencers (or wannabe influencers) are at risk of failing to meet their agreements with their sponsors and thus losing future earning opportunities.

Yet users do not enter pods only to increase their engagement. They also access them to trade information about the functioning of the platform with more experienced members. As Abramo, a twenty-nine-year-old Italian user, recalls: "Let's say that (a pod) is a sort of Instagram clandestine news bulletin, where people circulate news and tricks; then of course . . . if a news is being circulated by few groups, it's almost certainly a fake news, but if many people talk about that news, then it means that it might be true."

Thus, pods are both a "gym"—or an informal learning environment—where members train their knowledge on the platform, and a clandestine marketplace where its participants exchange valuable commodities ("likes," comments, and follow-backs).

TACTICS TO AVOID DETECTION BY INSTAGRAM

Two interviewees, Davide and Damiano, revealed to us a widely popular tactic to avoid detection by Instagram: when a pod member publishes a link to receive new interactions to their post, they delete the last part of the link to prevent Instagram from tracing it back to the platform on which the link was posted: "Many pods are banned because many people in the group share their posts leaving the link in full; this is something that we have understood over time," said Davide.

Editing the "tracking link" is just one of a set of clever tactics aimed at keeping a low profile to avoid ending up "on the radar of Instagram," as one of the interviewees told us. These actions consist in being careful not to exceed the daily limit of interactions allowed by the platform, or not to generate too many interactions in a short time because Instagram might understand this sudden increase in engagement to be inauthentic behavior fueled by a bot and consequently *shadow-ban* (block) the user.

Moreover, the members of a pod, just like the members of the underground society in the movie *Fight Club*, avoid talking to each other about the existence of the pod itself: "We don't talk about the pod in our chats with our friends; we have a little bit of fear . . . maybe you have a jealous friend who reports your Instagram profile and then Instagram puts you '*sotto torchio*';[57] so let's say that there is a bit of fear in saying you belong to a pod," Damiano remarked. The fact that he used the expression "*sotto torchio*" is a demonstration of how the members of the pod are aware of, and maybe even overestimate, the power of Instagram's automated patrolling system.

A "LIFE BELT IN THE SEA": THE MORAL ECONOMY OF PODS

From our fieldwork, we discovered that pod members are highly critical of those who resort to bots or that buy and sell "likes" and "follows." They consider these practices as forms of unfair competition and draw a clear line between the latter and their engagement communities. According to Stefania, for example, "pods guarantee a kind of growth that does not rely on any fake profile, but on real people with authentic profiles." Users know that pods are heavily discouraged by the platform, but they still frame them as a morally acceptable strategy because they consider them a fair and easily affordable defense weapon against the algorithmic government of visibility on Instagram. Joining an engagement group is considered an extremely useful tactics to start building an audience on Instagram. Abramo explained that, for a microinfluencer who wants to increase their visibility,

bots are not a good option, while pods I think it is the only way, if you can't afford paying for a post. With the fact that in Italy we are all too poor to pay for the sponsorship of a post, I think the pod is the only way to emerge, to start building our audience and get to a point where we can start earning money.

If an influencer has no money to invest in sponsoring their content, pods are considered the only way to increase visibility, but they are also perceived as a real life preserver. A microinfluencer explained his "addiction" to pods as follows: "I don't say that relying on these groups . . . becomes a drug, but they become the *lifebelt in the middle of the sea* . . . a user will always find it hard to abandon the pod, unless she reaches very high audience numbers; there are even some users who reached 50,000 followers and are still inside the groups." He also specified that unwillingness to abandon a pod comes from the fear of not being able to provide the service for which a company or a brand contracted him. Therefore, microinfluencers remain in the pod because "you never know. Maybe a big sponsor shows up and it's the chance of a lifetime; and then you ask yourself: Will I be able to reach the engagement rates that I promised him without the help of the pod?"

The pods are therefore safety nets, life buoys in the middle of the sea to cling to when someone risks a metaphorical drowning. This image evoked by one of the interviewees reminds us of the beginning of James Scott's book *The Moral Economy of the Peasant*, in which he compares the structural condition of the Vietnamese rural population to "that of a man standing permanently up to the neck in water, so that even a ripple is sufficient to drown him."[58] The comparison between an Instagram influencer and a Vietnamese peasant poised between death and subsistence may seem obscene, and it is if we understand it literally. Although a far cry from that of the Vietnamese peasants described by Scott, the comparison may be useful to eventually realize the precarious condition of Instagram's influencers and their extreme dependence on the force of the "tide" (i.e., the platform). The economy of the majority of influencers and independent content creators is also a subsistence economy, exposed not to climate change but to other, equally powerful, external forces such as algorithmic changes.

The majority of those we commonly call *influencers* are not public figures who wield enormous bargaining power against global brands and broadcast media. Most of them are precarious workers of the emerging cultural and creative industries,[59] or even Syrian refugees trying to make some tiny profit from livestream begging on TikTok, as the BBC discovered.[60] As Brooke Erin Duffy highlighted, while work in the media and cultural industries has long been considered precarious, the processes and logics of platformization have injected new sources of uncertainty into the creative labor economy: "Among the sources of such insecurity are platforms' algorithms, which structure the production, circulation, and consumption of cultural content in capricious, enigmatic, even biased ways."[61] The media scholar Zoe Glatt argued that, "contrary to highly celebratory discourses that position online content creation as more open and meritocratic than traditional cultural industries, this is an advertising-driven industry that propels the most profitable creators into the spotlight, resulting in the closing down of mobility."[62]

The visibility on which content creators build their reputation is volatile and unstable and depends on black-boxed algorithms. Rebecca, another microinfluencer, confirms that her job is extremely precarious and reliant on the visibility that she manages to attain. To represent an attractive investment for brands, she must constantly pursue quantifiable markers of visibility: "Unfortunately, sponsors require you have 5,000 followers at least, so you try to help yourself in some way, if you want to survive in this world." Damiano pointed out that "if sponsors see any drop in the engagement rates, they simply decide not to invest in that profile anymore." Some of the interviewees envision the pod as a form of retaliation "from below" toward the power of the platform. Instagram enforces its moral code through public announcements, community standards, and a continuous policing work over the platform, which aims to ban, or *deplatform*, those who do not respect the rules. Similarly, the pods have also set up their own methods to ensure that members respect the rules that they have set themselves. On Telegram, pod administrators use bots to verify that their members are doing what they promised, while on other platforms, members use apps to see which peers have stopped following them, and then undo their "likes" or stop following them in turn.

Most pod participants share a morally alternative code of conduct to the one enforced by the terms of service (ToS) of the platform. This code is made explicit by the rules set by each single pod and serves to establish what is deemed as acceptable and what is not. Similar codes have been observed also in engagement groups dedicated to Vinted or TikTok users.[63] In fact, this represents a real alternative "contract" to the Instagram community guidelines, as shown by the screenshots of the rules of some Italian pods (see figure 4.2, 4.3,[64] and 4.4[65]).

These screenshots show the concrete existence of primitive forms of everyday resistance to the moral economy of platforms like Instagram.

O 24 marzo 2018

A tutti i nuovi iscritti:
Prima di iniziare a postare, ricordatevi di seguire gli admin (che non sono tenuti a ricambiare)
Www.instagram.com/
Www.instagram.com/
Una volta fatto, commentate questo post con il link del vostro profilo instagram.
Chi farà unfollow agli admin verrà bannato dal gruppo senza preavviso e verranno eliminati tutti i suoi post!
Grazie!

To all new members:
Before starting to post, remember to follow the admin (which are not required to reciprocate)
Www.instagram.com/
Www.instagram.com/
Once done, comment on this post with the link of your instagram profile.
Who will do unfollow to the admin will be banned by the group without warning and all his posts will be deleted!
Thank you!

O 222   Commenti: 2105

4.2 Rules of an Italian Instagram pod on Facebook (with English translation).

In public, on the front stage, the members of the pod behave like common Instagram users, while once on the back stage, they perform actions that contrast with the rules of the platform. In these moral codes of conduct, so different from the community standards established by Instagram, we find material evidence of the existence of two contrasting moral economies—that of users and that of platforms—as theorized in chapter 2.

TWO COMPETING MORAL ECONOMIES

The moral economy of Instagram has been encoded into the terms of service and guidelines of the platform, while the rules established by pod members are the result of a normative process based on a diverse moral economy. Those who respect the pod's rules are committed to "growing with the community" and do not consider it immoral to resort to pods.

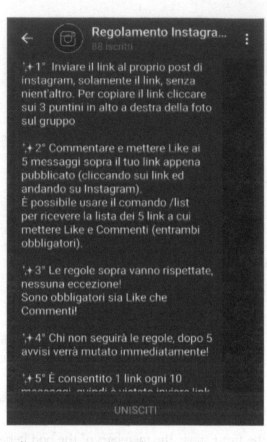

4.3 Rules of an Italian Instagram pod on Telegram.

The logic of the pod leverages on the ethics of collective growth: in the descriptions of the pods, the administrators talk about respecting the rules, "growing together," "reciprocate honestly," "don't be an asshole," "help," "give a hand," and "Please, reciprocate." What is this, if not the zero degree of a "mutualistic" language, based on a cooperative ethic, in which audience growth can only be achieved together? While the discursive regime of the pods highly values reciprocity and cooperation, Instagram emphasizes a narrative in which only the most talented are rewarded with visibility, a model of growth based on the individualistic and meritocratic morality typical of a neoliberal logic. Yet the moral economy of the platform is not entirely rejected by pod participants. The boundaries between the two moral economies are never clear-cut and

Informazioni su questo gruppo

Descrizione

pubblicate i vostri profili Instagram
chiedete scambio follow o like
ricambiate onestamente

● PER ESSERE INSERITI NEL GRUPPO SEGUITE I DUE
AMMINISTRATORI:

https://www.instagram.com/

https://www.instagram.com/   ●

🔒 privato
   Solo i membri possono vedere chi fa parte del gruppo e cosa pubblica

👁 Visibile
   Chiunque può trovare questo gruppo

👥 Generale

Membri · 276

4.4 Rules of an Italian Instagram pod on Facebook.

stable over time: pod participants often change their minds, argue that they want to leave the groups because they consider them unfair, but then return to them, maintaining a "double morality" for utilitarian purposes. Many of them unconsciously swing between one moral economy and the other.

ENTREPRENEURIAL SOLIDARITY UNDER CONSTRUCTION

Is the act of joining a pod related to self-interest, or does it imply being part of a community and feeling responsible for its success? There is no clear boundary between seeking personal interest or nurturing a sense of belonging to the community. We found various degrees of utilitarianism and mutualism among pod members. There are members who play the game and respect the rules only if the game provides them with personal benefits, while others develop a stronger sense of belonging to the group and invest a lot of time in sharing the skills that they have learned with the other members of the pod.

The sense of community is much stronger in topic-oriented or niche groups. Within these groups, our interviewees report that the exchange seems more "authentic and sincere." Groups with few members develop community ties more easily than those with thousands of members. As in the economy of gift described by Marcel Mauss,[66] the free exchange of "likes," comments, tricks, and tips could represent a form of "digital gift," which creates and maintains social ties and harnesses community members in a perpetual and binding circle of gifts and countergifts. Over time, this perpetual exchange of gifts and countergifts makes the members of the group more connected to each other, strengthening social bonds.

In generalist groups, on the other hand, the exchange of "likes" and comments often appears to the members as imposed from the top. Abramo told us that when someone leaves, in a few minutes all the pod members "unfollow" them. This immediate "punishment" has a rational explanation, as Abramo explained: "Why do I unfollow them? For an 'algorithmic' reason; because . . . I am afraid that, not belonging to this group anymore, they will become an inactive follower of my profile, and this will lower my engagement rate." The frailty and ephemerality of mutual support within an Instagram pod are strikingly reminiscent of the kind of solidarity that Scott found among Malaysian peasants. He wrote of an "imposed mutuality," a very fragile form of solidarity, which was hardly maintained through the imposition of rules and sanctions. As Scott illustrates:

Such minimal solidarity depends, here as elsewhere, not just on a seemly regard for one's fellows, but on the sanctions that the poor can bring to bear to keep one another in line. Since the temptation to break ranks is always alluring to members of a class that has chronic difficulty making ends meet, these sanctions must be powerful enough to prevent an ever immanent Hobbesian struggle among the poor. The modest level of restraint that has been achieved makes ample use of social sanctions such as gossip, character assassination, and public shunning.[67]

Similarly, even within pods, solidarity depends on "sanctions" imposed by pod administrators and is often at risk. In both thematic and generalist pods, a "flexible" and minimal solidarity emerges. It is characterized by a very low intensity of mutual care and of a contingent and volatile nature, which is based on an implicit contract of mutual aid and is imposed from the top by the administrators of the group through sanctions and rules. As soon as someone breaks the contract, the mutual aid lapses.

Many Instagram users joined a pod just to improve their visibility, but then, over time, they also developed a sense of belonging to the group and began to appreciate the solidarity among peers that they found within the pod. This solidarity, however, is too contingent and fragile to be mistaken for a form of group or class "consciousness." The solidarity that emerges from the practices of podders is quite similar to what Soriano and Cabañes call "entrepreneurial solidarity" (see chapters 2 and 3). This kind of solidarity deviates only partially from the moral economy of Instagram because while it challenges the platform rhetoric of competition and authenticity, it also helps to enhance content creators' entrepreneurial spirit. Finally, it lacks a "political" dimension that challenges the structures of power underlying platformized cultural industries.

The pods, in fact, are not the direct result of a preexistent "political" awareness of the pervasiveness of platform power. The decision to enter a pod does not depend on a preexistent sense of solidarity between influencers. If anything, the opposite is the case: it is only participation in the daily activities of the pods that can generate (more or less consciously) new bonds of solidarity, however volatile and "depoliticized" they might be. It is an entrepreneurial solidarity in perennial construction.

In the process of struggling against the affordances imposed by the platform, they establish themselves as a community of self-entrepreneurs. They enter the pod as individuals animated by self-interest (increasing the number of "likes"), but they may leave (or remain in) the pod as individuals that rediscovered the benefit of cooperation, mutual aid, and entrepreneurial solidarity. Solidarity is an ongoing process, not a given. Pods are only the first of many necessary steps toward building more structured forms of solidarity. The mutual support provided by the pods is aimed at fostering the survival of its members within the "like economy,"[68] not at changing its rules or building a more just and equitable cultural labor economy.

But the workers in the platformized cultural industries are not only organizing themselves in engagement groups, they are also building more structural forms of mutual support, such as new independent unions, a type of organization that is already emerging in all kinds of platform-mediated work.[69] In fact, new forms of labor organization are emerging also among freelance content creators working in the global influencer marketing industry. The YouTubers Union (YTU) was founded by a YouTuber named Jörg Sprave in March 2018, and in 2019, YTU teamed up with IG Metall, the largest trade union in Europe, to improve the working conditions of YouTubers who earn income, or aim to, from the videos they upload to the YouTube platform.[70] IG Metall, the German trade union, explained its decision to join the YTU by stating that several thousand people in Germany alone rely on advertising earnings from YouTube videos either as a main or important supplementary income source, and they represent a new frontier of struggles for workers' rights. Similarly, at the end of June 2020, the fashion blogger Nicole Ocran, and the influencer expert Kat Molesworth joined forces to launch The Creator Union (TCU), the UK's first union for digital content creators. The same month, an industry trade group called the American Influencer Council was launched in the US.[71]

Pods are places to incubate solidarity and awareness of platform power. In some cases, microinfluencers not only join a pod, but also resort to more structured forms of resistance, like independent unions. If being part of a pod can be understood as a tactic of everyday resistance, joining an independent union represents a stronger move toward forms of long-term, strategic agency that are not aligned to the moral economy of the platforms.

5

GAMING POLITICS

INTRODUCTION

Social movement activist Lucy is filming a police officer and live streaming their interaction on Instagram as legally allowed by the First Amendment. During their meeting, the police officer suddenly pulls out his smartphone and starts playing a famous pop song.

This surrealist scene could be the start of a David Lynch movie. Instead, it encapsulates the battlefield of contemporary algorithmic politics. The police officer is relying on a tactic that exploits social media's copyright protection algorithm to prevent himself from being filmed—and held accountable—by the activist. Any video that includes music, even just in the background, is potentially subject to removal by Instagram. An increasing number of political practices are being performed on algorithmically mediated environments, and different political actors have learned how to appropriate algorithms to fulfill their needs and objectives. This is happening in the context of a dramatic change in politics in the last decade. Digital platforms and algorithmic media are profoundly reshaping the ways in which political actors generate information, share their messages, and carry out their actions at the interstices between the digital and the physical realms. As this chapter will illustrate, at the center of this new type of algorithmic politics lies the quest for visibility and the use of algorithms to either artificially obtain it for oneself or to prevent others from gaining it for themselves. This aligns with many of the dynamics of the platform society that we have seen at play in the other chapters of this book. The current scenario appears as a multifaceted technopolitical battleground animated by different forces that coexist and clash, with multiple actors using algorithms to obtain control and enforce surveillance on one side, or for emancipatory aims in the pursuit of social justice on the other.[1] Within this contested battleground, platforms are also key actors with their attempts to moderate, channel, and control the flow of political information that circulates through their infrastructure.

In this chapter, we locate algorithmic politics within the broader scenario of data politics and draw a distinction between an *institutional/ strategic* type of algorithmic politics and a *contentious/tactical* one. We subsequently show the articulation between strategies and tactics that flesh out a taxonomy of three types of political engagement with algorithms (*amplification*, *evasion*, and *hijacking*) and discuss the moral economy of algorithmic activism. In the conclusion, we reflect on the key contributions of this chapter, interrogating connections and differences with the concept of hashtag activism and addressing the agnosticism of algorithmic activism.

THE EMERGENCE OF ALGORITHMIC POLITICS

The critical data scholars Evelyn Ruppert, Engin Isin and Didier Bigo have shown that data are not only shaping the relations within our societies but have important implications for politics and are inextricably linked to our democratic life. In sum, data create new power relations. These scholars argue that "data politics is concerned with the conditions of possibility of data that involve things . . . , language . . . , and people . . . that together create new worlds."[2] This politics consists in "both the articulation of political questions about these worlds and the ways in which they provoke subjects to govern themselves and others by making rights claims."[3] Data politics is thus "concerned with not only political struggles around data collection and its deployments, but how data is generative of new forms of power relations and politics at different and inter-connected scales."[4] Expanding on this definition and complementing it with the notion of "contentious politics" advanced by the social movement scholars Charles Tilly and Sidney Tarrow,[5] Davide Beraldo and Stefania Milan[6] have introduced a differentiation between the "institutional politics of data," which refers to the top-down effects of datafication on groups and individuals—and the "contentious politics of data," which denotes instead "the bottom-up practices embodied and promoted by individuals and groups."[7] With contentious politics of data, the two social scientists refer to "the multiplicity of bottom-up, transformative initiatives interfering with and/or hijacking dominant, top-down processes of datafication, by contesting existing power relations and narratives and/or by reappropriating data practices and infrastructure for purposes distinct from the intended."[8]

In the wake of this helpful distinction, we hereby propose a similar articulation for what we call *algorithmic politics*, a type of data politics that is concerned with how different groups exercise their algorithmic agency by appropriating and acting upon algorithms to fulfill their political objectives. Further, we differentiate between an *institutional* algorithmic politics and a *contentious* algorithmic politics. The former refers to the efforts to act on algorithms undertaken from above—by a state, an institution, a corporation, and so on. The latter represents instead all those practices initiated from below—by a collective, a social movement, a civil society organization, or individual activists. These two types of algorithmic politics connect two strands of literature that so far has rarely dialogued with each other. What we call *institutional* algorithmic politics is often referred to as "computational propaganda"[9] or "algorithmic populism,"[10] while the *contentious* algorithmic politics has been addressed as "algorithmic resistance"[11] or "algorithmic activism."[12] In addition, when analyzing these two types of algorithmic politics, we introduce a further axis of analysis based on de Certeau's distinction between the strategic and tactical dimensions that represents one of the conceptual pillars applied throughout this book. The introduction of this analytical dimension allows us to engage a fruitful dialogue with social movements and digital politics, while also speaking to the theme of resistance which relies at the core of this book.

INSTITUTIONAL AND STRATEGIC ALGORITHMIC POLITICS

Digital platforms, and social media in particular, play an increasingly central role in the dissemination and amplification of political voices. At the same time, social media represent environments where disinformation and hate have proliferated in recent years. The Wikileaks revelations, the Cambridge Analytica scandal, the Brexit campaign, the election of Donald Trump for US president, and the "infodemic" that accompanied the emergence of the COVID-19 pandemic are only some key events that have reignited debates on the role of digital technologies for the manufacturing of propaganda and manipulation.[13] The current conversation about the harms of digital platforms for democratic life has taken a decidedly darker turn following these and other similar occurrences around the world, and an idealized vision of the digital society "has been swiftly replaced by a recognition that our information ecosystem is now dangerously polluted and is dividing rather than connecting us."[14] Relatedly, new research has flourished along with a related conceptual vocabulary that either dusts off old notions and adapts them to the digital age (disinformation, misinformation, conspiracy theories, digital/online hate, etc.) or concocts new vague terms like "fake news"[15] and "post-truth"[16] to indicate that the scale of what we are experiencing is unprecedented. But these terms have not only been mobilized by scholars: the "fake news" moniker was frequently employed effectively by Trump to condemn any media coverage critical of him.[17]

To better account for the entirety of these various phenomena, researchers at the First Draft News organization have coined the term "information disorder"[18] to include fabricated, manipulated, imposter, and other misleading content, as well as false context and connection and even satire or parody when they are mobilized to spread rumors and conspiracies. Concomitantly, in the field of digital journalism, scholars like Thorsten Quandt have taken stock of the clash between the utopian, idealistic promise of the citizen participation ideal and the gloomy reality of organized hate campaigns, online misinformation, and the manufacturing of consent. Quandt has introduced the concept of "dark participation" to account for the variety of these antidemocratic forms of engagement in news production and circulation.[19]

We consider as both institutional and strategic all those manifestations of algorithmic politics enacted by institutions, governments, national states, corporations, think tanks, lobbies, public relations companies, and other entities that have social, economic, and cultural capital at their disposal to interfere with the work of algorithms, governed by long-term strategic visions.

For example, Ico Maly has put forward the notion of "algorithmic populism," which is "a populism that is at least partially reliant on applying algorithmic knowledge and activism in order to construct the populist voice and let it circulate throughout the hybrid media system."[20] He shows that nowadays, political discourse is significantly shaped by the algorithmic affordances of digital platforms. Populist voices are now able to exploit the media logic of the new hybrid media system[21] for the dissemination of propaganda and misinformation, as the architecture of social media seems to be particularly prone to these treacherous dynamics.[22] The monopoly of disinformation and manipulation is no longer the prerogative of a few powerful states and institutions. This kind of propaganda which relies on algorithmic power and operates through hybrid media environments, has been referred to as "computational propaganda."[23] This concept addresses the "use of algorithms, automation, and human curation to purposefully manage and distribute misleading information over social media networks."[24] Research in this field, for instance, has focused on the identification of political bots, also known as astroturf[25] accounts or sock puppets. Political bots are automated social media accounts that are built to look and act like real people and aim to manipulate public opinion.[26] Practices such as the deployment of political bots in relation to computational propaganda represent nonpassive forms of cohabitation with the work of platforms' algorithms, even if their aim is oriented to the strengthening of one's own political or economic hegemony. These kinds of algorithmic media are often mobilized to amplify and artificially boost the profiles of specific politicians, parties, and institutions to create an illusion of consent around controversial political positions, or to hinder political dissent and silence and fracture critical voices.[27]

While reflecting on the new dangers of institutional and strategic algorithmic politics, it is fundamental to avoid a technological deterministic tendency to overestimate the real power of algorithms and the role of social media in shaping politics and public life. The renewed interest in the most recent forms of algorithmically mediated propaganda has certainly helped to expose the extent of this phenomenon and the threats that these practices pose to democracy. Yet we should not embrace too linear conceptions of the relationships between technology, democracy, and society. As the global media and politics scholars Paula Chakravartty and Srirupa Roy have pointed out, in some of the recent academic studies on disinformation, populism, and propaganda, we are witnessing a "revival of the hypodermic syringe accounts of powerful media effects that we thought had been discredited by empirical evidence to the contrary."[28] The two scholars add that "these explanations for the rise of populism commonly reference the powers of media to persuade and ultimately distract or delude the voter from his or her 'real' concerns."[29] Hence, it is fundamental to look not only at the computational side of these processes, but also at their political dimension. Research on institutional algorithmic politics need to engage with the power relations involved and the motivations behind the social actors who sustain, promote and benefit from these practices.[30] What we are witnessing is a political and epistemic crisis, and the role of technologies within it should not be overhyped but rather critically assessed, including the conditions that can contribute to fuel or mitigate this crisis.[31]

Looking at the people behind the dynamics of strategic algorithmic politics means also considering the workforce behind digital propaganda and manipulation and recognizing the motivations that drive several people, especially in the Global South, to partake in this business. The ramifications of these strategies are global and have unfolded not just in the Western world, but in countries of the Global South as diverse as the Philippines[32] and Mexico.[33] "Troll armies" and "troll factories" are increasingly deployed worldwide to spread disinformation, manufacture consent, undermine protest, and interfere in political opinions and decision-making processes. In this context, it is important to zoom in on the production mechanisms of digital disinformation, adopting a perspective that is close to that of media production studies. We should avoid the tendency to portray trolls with stereotypes such as that of "folk devils,"[34] responsible of igniting moral panics in society. Adopting a perspective that focuses on the agency of the people who produce this content is key to get beyond stereotypes about algorithmically mediated' forms of control.[35] The media scholars Jonathan Ong and Jason Vincent Cabañes[36] were among the first to propose a dialogue between media production studies and disinformation studies to illuminate the complex sociotechnical network of actors and infrastructures that lies at the heart of the artificial shaping of politics. The two scholars highlight that it is necessary to locate troll farms and whom they call the "architects of disinformation" not outside the media industry but, on the contrary, firmly "within broader media ecologies."[37] This allows us to understand paid trolls not as irrational and anonymous folk devils, but as precarious, underpaid workers who are part of their domestic media industries and key actors of our hybrid media systems.

While most of the scholarly attention has been devoted to exploring algorithmic governance and top-down forms of digital politics and control from established and institutional actors, social movements and activists around the world are also appropriating algorithmic power in their attempts to transform society and bring about social change. In the next section, we tackle contentious and tactical forms of algorithmic politics focusing on algorithmic activism.

CONTENTIOUS AND TACTICAL ALGORITHMIC POLITICS (ALGORITHMIC ACTIVISM)

Collective action is a powerful force that drive social change in our societies. Throughout history, protest movements have induced major societal, cultural, and political shifts. Exploring their tactics is key to illuminating how algorithms are envisioned and enacted in the struggle for a more just society or governed by values other than hegemonic ones. Social movements are also central to our understanding of innovative technological appropriations and experimentations. Across the ages, activists have proven how to successfully deploy technologies in ways that differ from the intentions of their creators to fulfill their needs and political objectives. Relying on their creativity and dealing with a scarcity of resources, they have been pioneers in the creation of independent media infrastructures and the subversion of corporate digital platforms.

At the same time, social movements have constituted privileged settings for the development and diffusion of radical and counterhegemonic social imaginaries. Activist formations represent spaces where different ways of thinking about democracy, equality, and justice materialize and where diverse ways of using technology are cultivated and enacted. Algorithmically mediated environments are restructuring collective action and the dynamics of social movements at a profound level[38] changing the very conditions under which social movements operate.[39]

Since most digital activist practices now reside within the contours of corporate digital platforms, activists must deal daily with the complexities of algorithms to promote and carry out their protest activities. Algorithms have several implications for the activities of social movements and represent key actors in what German sociologist Ulrich Dolata has called the "socio-technical constitution of collective action."[40] But contemporary activist formations show us that social actors can resist, subvert, and repurpose algorithms to envision alternative social imaginaries and foster different kinds of data futures.[41] While data studies have paid much attention to the dystopian consequences of data systems and regimes of algorithmic governance, examinations of activist engagements, appropriations, and experimentations with algorithms are still rare. The data scholar Helen Kennedy has pointed out that activism that engages with data and algorithms "requires the possibility of agency, yet there is little scope for agentic engagements with data in the visions of datafication provided in much data studies scholarship."[42]

As we have illustrated throughout this book, in order to grasp how algorithms can be repurposed and reconfigured, we need to look at the agency of the social actors involved, intended as "the ability of social actors to variably engage with, and react to, the context in which they are embedded, that empowers them to change their relation to structure."[43] Algorithmic agency, intended as the reflexive ability to make algorithms work to their own needs (as discussed in chapter 1), does not exclusively reside in the hands of elites, institutions, governments, and corporations. As this book has demonstrated, this agency can be exercised from above and below by various actors, and for contrasting purposes. Social movements, civil society organizations, community groups, and numerous grassroots and alternative actors can appropriate and repurpose algorithms in counterhegemonic ways to advance social justice and to imagine and unfold different "data worlds."[44] Studies on data agency,[45] data activism,[46] and everyday practices of "living with data"[47] concur in underlining that the analysis of top-down processes of datafication is just one side of a more contested battlefield. Activists can make sense of algorithms from below and repurpose them to pursue social transformation in their efforts to change our societies.

Following this line of thought, Maly[48] has devised the concept of "algorithmic activism" to tackle the rise of a Flemish far-right activist movement. He sheds light on how a new breed of algorithmic activists from this movement strategically exploit the inherent affordances of social media to reach their goals, "boost their popularity rankings,"[49] and make their content go viral. In this chapter, we identify *algorithmic activism*[50] with the contentious politics of data that is concerned with how a range of actors including activists, social movements, and civil society actors engage with, and act upon algorithms to achieve their political aims and pursue social change. We therefore conceptualize this phenomenon as a form of *contentious* algorithmic politics exercised through a *tactical* form of algorithmic agency. As we will see—and as Maly's example illustrates—algorithmic activists can also tactically leverage the algorithmic opportunities of digital platforms to push conservative, racist, and right-wing agendas. We refer to this aspect as the *agnosticism* of algorithmic activism in the conclusion of this chapter.

It is essential here to recall the articulation between "algorithms as stakes" and "algorithms as repertoire" that was outlined in chapter 1, drawing on the distinction of Beraldo and Milan between "data-as-*stakes*" and "data-as-*repertoire*." [51] In the former type (*data-oriented* activism), data are the "main stake in a hypothetical claim-making agenda."[52] In the latter (*data-enabled* activism), they are instead incorporated within the repertoire of action of social movements and activists,[53] "alongside other more traditional forms of protest and civic engagement."[54] Applying this analytical articulation to the realm of algorithmic activism, we can foreground a type of *algorithm-oriented* activism that sheds light on negative effects of platform power, and another type of *algorithm-enabled* activism that instead deploys algorithms as repertoire. The former kind denounces the several biases that are reproduced by the proprietary algorithms of digital platforms. Studies on algorithmic bias, oppression and discrimination have made a strong contribution to our understanding of the potentially devastating effects that algorithms can have on our society, democracy, and culture.

The work of such scholars as Virginia Eubanks,[55] Sofya Noble,[56] Ruha Benjamin,[57] and Sasha Costanza-Chock[58] represents a turning point for critical data and algorithm studies. These thinkers raised awareness among citizens and global civil society about the many perils of an algorithmically governed society and have confronted the problematic assumptions and decisions in relation to race, gender, status, class, and the various forms of domination and discrimination that are encoded, designed, perpetuated, and intensified by algorithmic systems. They have also proposed remedies to the systemic causes and the structures of oppression that circumscribe the effects of algorithms on society. The British computer scientist Dan McQuillan, for example, argues that we need to develop an "antifascist" artificial intelligence (AI).[59] He means that we need to do more than debiasing data sets, democratizing participation to the engineering elite, or rendering AI more "ethic." McQuillan argues that this approach must be both decolonial and feminist and proposes to create structures where those affected by AI discrimination can contest machine decisions through the collective refusal of automaticity.

At the same time, a growing number of initiatives and organizations[60] are focusing on data and algorithms as *stakes*, developing forms of activism that highlight the harmful effects of platform power and the inequalities that are often reproduced by data systems and the proprietary algorithms of digital platforms. This led to a questioning of the deployment of algorithmic systems at a structural level in ways that oppose the inevitability of the adoption of these technological developments.

While these two types of algorithmic activism are inextricably connected (as illustrated in chapter 6), the key focus of this chapter lies on the latter type of algorithmic activism, where algorithms are integrated into the *contentious repertoire* of social movements and civil society actors to pursue political transformation. The social movement scholar Charles Tilly originally defined a "repertoire of contention" as including the "whole set of means [that a group] has for making claims of different types on different individuals."[61] He later clarified that this notion refers to "claims making routines that apply to the same claimant-object pairs: bosses and workers, peasants and landlords, rival nationalist factions, and many more."[62] More recently,[63] Tilly extended his focus to include broader, contentious performances, emphasizing the constant innovation of contentious politics. Reflecting on the increasing relevance that new media have obtained within the repertoires of social movements, scholars have started to develop notions such as "electronic repertoire of contention"[64] and "digital network repertoires."[65]

Both social movement and media scholars have adopted the concept of repertoires of contention to study how activists integrate digital tools as means to challenge authority, protest, and mobilize.[66] Algorithms exemplify the latest addition to the repertoire of contention of social movements, since the capacity to understand, adapt and repurpose algorithms lies at the center of today's collective action. Different thinkers have touched upon this topic in disciplines as varied as computer science, political science, media, data, algorithm, management, resistance, and social movement studies. For example, scholars working at the intersection between social movement, media, and algorithm studies have addressed the ways in which activists and algorithms are mutually enmeshed in protest-related settings. In their ethnographic account of digital protest in Greece and Sweden, Vasilis Galis and Christina Neumayer[67] have introduced the concept of "cyber-material détournement" to indicate the alliances between activists and algorithms that define social media activism. Other theorists have grown dissatisfied with dystopian accounts of algorithmic power and started to rely on de Certeau's understanding of tactics. Julia Velkova and Anne Kaun, for instance, "foreground the significance of mundane user encounters with algorithms through which users can develop tactics of resistance through alternative uses."[68] Relatedly, such experts as Justine Gangneux use the term "tactical agency" to designate the ways in which young people engage (and disengage) with WhatsApp and Facebook Messenger.[69] Tanya Kant[70] describes the practices of social actors who are engaged in "manoeuvring within, against and through algorithmic anticipation"[71] as "algorithmic tacticians."

Activists engage in algorithmic activism by leveraging the affordances of social media's algorithms to various ends. Mostly, their goal is to obtain mainstream media's coverage to increase their visibility (for the relevance of visibility, see chapter 4). In this context, visibility can be understood as the "digital embodiment and online presence of individuals and groups and their associated meanings, which are . . . constantly negotiated, reinvigorated, and updated."[72] Algorithmic visibility is a new kind of power[73] that selects which actors and which content can or cannot be seen. In algorithmically infused societies,[74] it is essential for activists to play the game of visibility.[75] This kind of digital visibility is also deeply ambivalent: for civil society actors, being in the spotlight can mean recognition and empowerment, but also being exposed to more surveillance and control.[76] The dual nature of algorithmic visibility is exploited by activists themselves, as we will see in the next sections. Yet campaigners also appropriate algorithms to unfold their "narrative agency"[77]; that is, the capacity to tell their stories and frame actions and experiences in their own terms. This is testified by movements across the world as diverse as #OccupyWallStreet, #BlackLivesMatter, and #YoSoy132. As the social movement scholar Zeynep Tufekci has pointed out,[78] the most successful movements are the ones that can develop a "narrative capacity," intended as the ability to attract public attention and insert new issues and frames into the political debate. In the platform society, algorithmic activism represents a central element of the formation of this narrative capacity for activist collectives.

A TAXONOMY OF ALGORITHMIC POLITICS

In this section, we illustrate the variety of practices in which algorithms are used as repertoire by a variety of both institutional and contentious actors around the world. We show how different types of algorithmic politics center on a struggle around voice and the incessant pursuit of visibility. This entails attempts to achieve that a specific political entity's voice is recognized, and its visibility strengthened, while at the same time, it includes the negation of other social actors' visibility and the attempts to silence their voices. We present a taxonomy of three types of algorithmic politics: *algorithmic amplification, algorithmic evasion,* and *algorithmic hijacking.* It is important to clarify that these categories represent ideal types of algorithmic practices that in reality are much more enmeshed and interconnected among each other. We have decided to single them out to better foreground the key dynamics at play and the incessant back-and-forth struggle between algorithmic strategies and tactics that characterize contemporary digital politics. As usual, our focus is lying predominantly on the tactics developed from below: for this reason, we provide ample space to chart the dynamics of algorithmic activism.

ALGORITHMIC AMPLIFICATION AND THE MORAL ECONOMY OF ALGORITHMIC ACTIVISM

In specific political junctures, both institutional actors and different kinds of activists can integrate algorithms into their repertoires to artificially multiply and amplify their voices and acquire more visibility. For example, the "manufacturing of consensus" is now an integral part of the toolbox of institutional digital politics across many countries of the world,[79] as evidenced in political contexts as diverse as the US,[80] Mexico,[81] Kenya,[82] and Argentina.[83] The aim is to employ automation to artificially boost the visibility of and generate an inflated image of popularity for some political candidates, a process that has been described in reference to Argentinian politics as being "popular with robots."[84] The perceived popularity of these candidates can generate a "bandwagon effect"[85] that motivate other people to "follow" and "like" them on social media platforms, thus further increasing their algorithmic visibility and the related illusion of popularity.

The algorithmic amplification of visibility and popularity represents a clear issue for democracies, as it stifles and poisons the possibility of genuine debate and the authentic process of participation that are required for democratic societies to function. As the political scientists Tobias Keller and Ulrike Klinger point out in their assessment of bots in election campaigns, "the principle of plurality is based on the premise of authentic interests and stakes in a society. Bots may insert non-authentic interests (interests no human or group in a society has ever voiced) and manipulated interests (fake interests that are manufactured to distort plurality). It becomes impossible for a society to monitor itself when machines disguised as societal members enter and manipulate the marketplace of ideas."[86]

Activists too have demonstrated the ability to effectively integrate algorithms into their repertoire to multiply and amplify their voice to acquire more visibility while strengthening their agency and narrative capacity. For example, political science scholars have addressed the use of algorithms on Twitter to pursue social change and insert alternative narratives through a notion called "hashtag activism."[87] These researchers examine how marginalized groups rely on Twitter hashtags (e.g., #MeToo, #GirlsLikeUs, and #BlackLivesMatter) to preempt political spin, erect networks of dissent, and challenge hegemonic understandings of gender and race. In our own previous work,[88] we devised the notion of "algorithmic resistance" to characterize various tactics of appropriation and repurposing of social media algorithms by social movements to pursue their political aims and achieve greater visibility. This concept originates from extended fieldwork into the technopolitical practices of the Spanish movement 15M (also known as "The Indignados"). Through the tactical adoption of social media and their algorithms, this movement was able to spread information, organize protests, build powerful narratives, and shape both national and global media coverage.

One of the most representative tactics of algorithmic activism established by these activists consisted in the systematic creation of trending topics on Twitter.[89] This tactic can be conceived as a communicative practice that encompassed a meticulous combination of internal communication technologies and dissemination on social media platforms. Spanish activists used tools such as online pads (digital notepads for collective writing) to (in the words of an activist) "collectively select a list of possibly successful hashtags and build the narrative of the protest." Then, public-facing environments like social media (primarily Twitter) were used to massively spread the information and reach the desired results. On notepads, activists would brainstorm and debate the most effective hashtags and then pick some of them depending on the activist actions that needed to be promoted. Once a hashtag was selected, a range of potential tweets was created accordingly and shared with other campaigners through internal communication tools. Also, 15M activists used a variable mixture of instant messaging services (WhatsApp, Telegram, and Signal), text messages, traditional mailing lists, and direct messages on Twitter and Facebook Messenger. This advanced tactic of algorithmic amplification presupposes a deep awareness and the understanding regarding the ways in which social media algorithms operate. As one

15M activist clarifies, this awareness was "obtained through incessant sequences of try and error," originating from "trying to understand how the Twitter algorithm worked and how we could exploit it for improving our visibility, spreading our activities and influence the mainstream media agenda."

As another Spanish activist further elaborates:

Our aim was to hack the Twitter algorithm so our narratives, our voices, our ways of seeing things could reach as many people as possible. We use these corporate social media because this is where most people are, and we want to reach the highest number of persons out there. We want to be visible, and we want our message to get across and be picked up by other mainstream media as well. We know that the nature of commercial social media is extractive and that they use our data in many ways, but this time it's us using them to multiply our presence, spread our messages and empower our movement.

Through their continuous technological endeavors, social movement actors found that general trending topics had a cycle of twenty-four hours and "all the accounts needed to tweet simultaneously with the same hashtag" (interview with a media activist). Furthermore, they detected that the hashtag had to be "fresh" because the Twitter algorithm always rewards newness. It is telling that in the quote given here, the Spanish activist specifies that that "the nature of commercial social media is extractive" in relation to activists' data, and tactics of algorithmic amplification that empower the movement are a way to use them back. This points to the existence of an alternative moral economy that justifies the artificial creation of popularity from below because corporate digital platforms are often seen as commercial predators with no social values and no scruples, and thus they are conceived as environments that can be exploited by any means necessary to advance real democracy and social justice causes. Another illuminating quote from an experienced hacker also touches upon this issue:

We are very aware that social media are toxic spaces of data extraction. Besides, being on them expose you to control and surveillance by authorities and the police. But we are not naïve either . . . For the first time, we have the knowledge to utilize them in ways that can massively multiply our message and discourses and reach a lot of people . . . Because this is where the people are, we need to be there, and we have the right to strengthen a message that promotes democracy, equality, and real participation.

This last quote also aligns with what the Italian sociologist Paolo Gerbaudo has called the move from cyberautonomism—that characterized the global justice movement and the creation of alternative media like Indymedia—to cyberpopulism[90] that defines contemporary activism. In the latter, movements do not shy away from commercial platforms and adopt a pragmatist attitude in the gaming of their algorithms[91].

A similar tactic has been observed by Halperin[92] on Facebook, where left-wing groups have developed tactics of "counter-populist algorithmic activism" to increase the visibility of their comments. This consists, for instance, in sharing popular posts by right-wing politicians on Facebook with the group's followers. Then, a subset of the group members would visit the post leaving critical comments, and other members would return to their group's page to share links to the antagonistic comments that they posted. As Halperin explains, "Once such comments are in place, a band of around 200 to 300 members who are available at the time follow these links *en masse* and "like" the messages left by their peers."[93] Facebook's ranking algorithm relies on the principle of relevance,[94] which is determined by levels of engagement (e.g., the numbers of "likes" to a comment). Consequently, the first thing that many ordinary social media users who visit these posts would encounter is the critical comments from leftist activists. This form of resistance exploits Facebook's ranking algorithm: the group aims to boost the visibility of members' comments on popular Facebook pages "to restructure the manner in which the public make its appearance on social media and demonstrate to online audiences that large parts of the people vehemently reject the Right's exclusionary populist vision."[95]

In other cases, activists rely on algorithms to amplify the visibility of someone in order to expose them with the aim of hurting their public image. In 2018, in response to Donald Trump's policies on immigrations and lesbian, gay, bisexual, transsexual, and queer/questioning (LGBTQ) rights, activists started to manipulate Google's search algorithm by massively linking the word "idiot" to pictures of him.[96] The association between the word "idiot" and the president of the US at the time was partly ignited by the choice of London protesters of the Green Day song "American Idiot" during Trump's visit to England. Activists on Reddit started to upvote posts of Trump[97] associated with the word "idiot," leading Google's ranking algorithm to establish that connection. Google has a long history of this type of scandal, with offensive and sometimes racist content being linked to specific individuals: in 2009, searches for "Michelle Obama" returned a picture of the first lady's face with apelike features. To date, the company's stance in relation to this and similar incidents has been not to interfere with its research results. These controversies highlight the inherent biases of algorithmically driven decisions,[98] but they also exemplify the changing and contested dynamics of algorithmic power where different actors (digital platforms, political figures, activists, the public, etc.) and logics interact.

Finally, activists can integrate algorithms into their repertoire to respond to these forms of algorithmic injustices. For example, Julia Velkova and Anne Kaun[99] have focused on forms of explicit algorithmic resistance though their concept of "media repair practices." These "media practices of repair are tactics to correct existing shortcomings *within* algorithmic culture rather than by producing alternative pathways. In that sense, they establish reactive user agency in an algorithmic aftermath."[100] The two scholars investigate the World White Web project by the Swedish design student Johanna Burai, who sought to tweak Google's image search algorithm to amplify alternative search results after discovering that her basic photographic search of a human hand returned almost exclusively images of White hands. In 2015, Burai launched a campaign to get six images of nonwhite hands to be among the top results in Google image searches. Practices of algorithmic amplification show that activists can appropriate algorithms to achieve more visibility or amplify someone else's biased voice to hurt their public image. This latter case displays how protesters can repurpose algorithmic power to repair what are perceived as unjust forms of invisibility.

ALGORITHMIC EVASION

This category includes examples of various social actors fighting against digital platforms' codes and algorithmic regulations to have their voices heard and their messages spread. This includes the realm of "content moderation avoidance strategies." These strategies have been observed in spheres as varied as vaccine-opposed groups,[101] pro–eating disorder (Pro-ED) communities,[102] and far-right movements.[103] In relation to the first type, Moran et al.[104] have exemplified how COVID-19 vaccine-opposed groups are able to bypass community guidelines and moderation features on social media through various maneuvers. One of the most diffuse of these techniques consists in lexical variations on Twitter, where iterations of words (e.g., "v@ccine") are used to prevent algorithmic detection and the blocking of content. On the same platform, they would also create vaccine-neutral hashtags to insinuate misinformation into pro-vax conversations. The practices of antivax strategists, these scholars show, are always carefully tailored to the specific platform architecture that they are appropriating. On Instagram, they would foster ephemeral content strategies including antivaccination content on stories rather than in-feed, or rely on coded language with words like "toxins" and "metals" instead of "vaccines." On Facebook, they would provide links to vaccination misinformation in the comment section rather than in the original post.

In the case of Pro-ED groups on Instagram, similar strategies of content moderation circumvention were identified in response to Instagram's ban of phrases such as "anorexia," "proana (pro-anorexia)," "thinspiration," "thighgap," and "imugly," as well as the hindering of the results for certain hashtag searches since 2012. This has led to a multiplication of lexical variants of the banned tags to deceive algorithmic controls. Pro-ED would replace "thighgap" with "thyghgapp" and "thinspo" with "thinspooooo." Research[105] has demonstrated that the Instagram ban exacerbated the situation since new variants obtained an even further reach than those that they were designed to replace, spreading to other platforms such as Tumblr and Twitter.

Relatedly, Prashanth Bhat and Ofra Klein[106] have elucidated how far-right activists evaded censorship by Twitter's automated moderation algorithm during the 2016 US presidential election. These scholars draw on the notion of "dog whistling" to describe the use of symbols and terminology that means something to the larger public but acquire a different meaning for a more specific group. In this case, words as "googles" or "Dindu Nuffins" were used to indicate African Americans and "skittles" to designate Muslims for a white supremacist audience. Other symbols, such as parentheses and percentages, were used to spread hate and racist, xenophobic content that escaped Twitter's algorithmic filters. The reliance on a particular language for avoiding algorithmic control and surveillance has been also observed in contemporary Turkey, but in the hands of activists opposing the authoritarian government. Tønnesen[107] underlined how activists who oppose the government appropriated the "vernacular Twitter language and the Western popular culture, which includes drawing subtle comparisons between movie villains and the President, references to early Internet phenomena, captioning viral videos with implied political messages etc."[108] This appropriation allows them to fly under the Turkish government's social media control while still getting their message across; this playful vernacular language is out of sync with progovernment forces' digital tools. Other examples of evasion tactics include far-right Italian activists on Twitter using symbols and numbers instead of letters (see figure 5.1 and 5.2) when writing highly critical tweets targeting specific politicians. This is

5.1 Tweet from an Italian right-wing activist on March 18, 2022, commenting on the views of a neoliberal Italian politician, Michele Boldrin, accusing him of being a "deranged" person: the encrypted message stands for "Boldrin squilibrato" ("Boldrin deranged").[110]

5.2 Tweet from an Italian far right-wing activist on March 19, 2022. The cipher stands for "Riccardo Bauer è un povero coglione" ("Riccardo Bauer is a poor asshole"). Riccardo Bauer (1896–1982) was an Italian antifascist politician.

carried out to cheat Twitter's algorithm and avoid censorship while at the same time creating a sense of community with other likeminded users.[109]

Finally, algorithmic evasion includes tactics of obfuscation and disengagement. The former has been conceived as a form of vernacular resistance to the surveillance regime of digital platforms.[111] As Brunton and Nissembaum articulate, "With a variety of possible motivations, actors engage in obfuscation by producing misleading, false, or ambiguous data with the intention of confusing an adversary or simply adding to the time or cost of separating bad data from good."[112] From radar chaff to BitTorrent and TOR, people can rely on a plethora of anonymizing technologies to protect their activities and identity from algorithmic control. In the latter case, algorithmic evasion might involve political disengagement, as people "may stop acting politically on social media platforms as a way of avoiding an algorithmic visibility regime that is felt as demeaning their civic voices."[113, p. 77]

Based on his research in Brazil, João Magalhães explores how citizens disengage from Facebook because they perceive that the existence of algorithmic bubbles harms their citizenship. Moreover, they feel that the cost of visibility is too high, entailing "unacceptable sacrifices to their values and emotional well-being"[114] and causing distress because of the impossibility to fully control their visibility even when it is achieved. This experience illuminates the moral economy of algorithmic activists from a different point of view. Both 15M activists and Brazilian users share a vision of social media as extractive and toxic commercial platforms. However, whereas Spanish protesters engaged in amplification tactics to (as the activists we interviewed would tell us) "hack these environments" for their own causes, Brazilian users abstain from social media. These cases prove that algorithmic activism is not only about appropriating algorithms to amplify visibility and strengthen narrative agency, but also about finding new ways to evade them to avoid being silenced, tracked, or even recognized by digital platforms.

ALGORITHMIC HIJACKING

With the term "hashtag hijacking," computer science literature defines a practice where hashtags are used to spread unrelated, negative content or spam. The goal is to tarnish the intended motive of a hashtag, thus rendering its presence counterproductive.[115] The following case study from Mexico will elucidate this dynamic. On September 26, 2014, six deaths and the forced disappearance of forty-three students at the Ayotzinapa teachers' college in Guerrero spurred the emergence of a social movement in solidarity with the families of the victims, whose main aim was "to present the missing students' lives."[116] The spark of the movement was ignited after the event, when outraged activists started to protest on social media, creating the Twitter hashtag #YaMeCanse (IAmTired) that expressed the feeling of not being able to stand any more violence. The hashtag soon became a key space for protest organizing, information spreading, and one of the most successful hashtags in Mexican history. The multimedia artist and writer Erin Gallagher has collected a detailed database of bot attacks and algorithmic strategies in Mexico and beyond.[117] In relation to #YaMeCanse, she noticed something atypical in her search results for these hashtags: they were flooded with tweets that had no content besides random punctuation marks. The accounts tweeting this empty content were in fact bots with no followers, which were tweeting automatically. As documented by the Mexican activist and blogger Alberto Escorcia,[118] automated accounts purposively hijacked the hashtag inserting "noise" through links to pornography, advertisements, and violent images. The Twitter algorithm would flag it as spam and consequently block it. The confusion that this generated made it challenging for citizens to share information using hashtags that vanished from Twitter's list of trending topics. However, activists reacted to the attempted hijacking of their hashtags by applying the previously described tactic of algorithmic evasion: they produced different lexical iterations of their original hashtag, such as #YaMeCanse1 and #YaMeCanse2. These iterations were able to move the conversation elsewhere, escaping the confusion created by the government's automated trolls.

This case exposes the battlefield of algorithmic politics with activists engaged in tactics of algorithmic evasion and reamplification in response to top-down strategies of algorithmic hijacking. Alberto Escorcia is part of a new generation of activists who have been detailing the rise of bots, trolls, and fake profiles in Mexico since the 2012 elections and publishing his original analysis of hashtags, trends, and data on his platform LoQueSigue (WhatFollows).[119] In this blog, the Mexican activist has documented the use of automated accounts to disrupt protests in numerous political campaigns. Relying on social network visualization tools such as Flocker and Gephi, Escorcia has developed ways to detect bots by examining the number of connections that a Twitter account has with other users. He also creates videos explaining to the public the dangers of bots and algorithmic attacks, proposing recommendations to help activists to counteract strategies of algorithmic hijacking. As he illustrated during an interview with us:

You should always post new, fresh content and avoid posting the same stuff all over again because Twitter's algorithm favors novelty. Also, you should devote time and resources to build real and strong connections in your activist network to show to the algorithm that you are not in fact a bot. Moreover, as it happened with YaMeCanse, you can build iterative versions of the hashtag with numbers like "YaMeCanse1," "YaMeCanse2," etc. to avoid bot armies that are trying to drown out real conversations with noise. In that way, you can move the discussion elsewhere and continue.

This example shows how institutions can deploy strategies of algorithmic hijacking to prevent activists' voice from being heard and silence online dissent. Yet activists can quickly use algorithms to react to these strategies. Through algorithmic evasion and amplification, activists reclaim spaces for their own narratives in the never-ending confrontation between strategies and tactics that defines contemporary politics. Another illustration of this tactic is represented by the collective hijacking of the Twitter hashtag #myNYPD described by Sarah Jackson and Brooke Foucault Welles,[120] that followed the launch of a public relations campaign by the New York City Police Department in 2014. The scholars document how activists' hijacking of the police hashtag led to the formation of a counterpublic sphere. They rely on this experience to demonstrate how the activists' repertoire of contention is evolving in the algorithmic age.

K-pop fandoms—known for their dedication to the idols of South Korean music—have recently received journalistic coverage for their algorithmic skills at the service of social justice causes (see also chapter 4). In 2020, protests emerged in response to the police killing of George Floyd in Minneapolis, and the subsequent threat of President Trump to employ the army against demonstrators. In this scenario, the Dallas Police Department asked the public to submit video of "illegal activity from the protests" through a dedicate app. K-pop fans inundated the app with K-pop-related content such as "fancams" (fan-edited videos of K-pop stars). As a result, the Dallas Police Department was forced to remove the app.[121] K-pop fans are also responsible for hijacking right-wing hashtags like #MAGA, #BlueLivesMatter, and #whitelivesmatter, flooding them with clips of K-pop groups, memes, and similar content to drown out racist and other offensive voices. As the digital culture researchers Crystal Abidin and Thomas Baudinette[122] have remarked, the fact that K-pop fans are mastering these tactics is no surprise to fandom scholars who have been following their practices for years. These communities have a long history of "being political," using online platforms to promote charitable causes, online vigilantism, and social justice goals.[123] Through their constant engagement with the architecture of the social media ecosystem, they have developed a strong algorithmic awareness and refined tech-savviness.

Finally, algorithmic hijacking is not merely related to hashtags; it extends to other devices and practices. For example, police officers in the US (see the sketch at the start of this chapter) have been playing copyrighted music with their phones during attempts to be recorded by activists in the course of their encounters. Their aim is to prevent those videos to be posted on YouTube, thus eluding accountability. To do so, they exploit the platform's copyright policy, which filters and removes this kind of material through a system called Content ID.[124] This kind of algorithmic hijacking to silence activists' voices illustrates the hybrid nature of the phenomenon that can manifest at the intersection between the physical and the digital spheres.

BEYOND HASHTAG ACTIVISM: FINAL REFLECTIONS AND KEY TAKEAWAYS

This chapter has represented a journey into algorithmic politics, an area of inquiry concerned with how different groups appropriate algorithms to meet their political objectives. We have introduced the distinction between an institutional and strategic use of algorithms for political purposes and a contentious and tactical one. This second type epitomizes the field of algorithmic activism. Within this kind of activism, algorithms can be the object *against* which activists' actions are oriented (algorithms as *stakes*) or the tool *through* which they exercise their protest (algorithms as *repertoire*). We have shown that algorithms represent the latest addition to the repertoire of contention of social movements, and we have discussed a taxonomy of three types of algorithmic politics (*algorithmic amplification*, *algorithmic evasion*, and *algorithmic hijacking*) to illuminate the actors and forces at play within this technopolitical battleground and reflect on the moral economies sustaining algorithmic activism.

It has become clear that our conception of algorithmic activism includes but exceeds the notion of "hashtag activism." While Twitter has been a fundamental space for recent forms of contention, algorithmic activism encompasses a wider range of tactics, dynamics, and stratagems and navigates a broader ecosystem of platforms, devices, actors, and sociopolitical territories. Moreover, as our book clearly demonstrates, algorithms are gradually pervading more aspects of our lives, and thus these practices are becoming more mundane. It is also key to underline the *agnosticism* of algorithmic activism. As we have shown, algorithms are being appropriated both by right-wing, racist, and misogynistic movements and by oppressive and authoritarian regimes. At the same time, they are employed by progressive activists, left-wing movements, and radical collectives. Hence, the battlefield of algorithmic politics and the fight for visibility is multidimensional and ambivalent.[125] It is defined as an incessant struggle between strategies and tactics and as disruptions and alliances between various actors, with platforms playing an increasingly salient part in this game. It is also defined as competing moral economies, with activists reacting to the extractive nature of commercial platforms through appropriation and engagement or choosing instead forms of withdrawal and abstention.

More than ten years ago, the political theorist Jodi Dean argued that "globally networked communications remain the very tools and terrains of struggle, making political change more difficult—and more necessary—than ever before."[126] These words resonate stronger than ever in the platform society, where political change and algorithmic agency continuously intersect.

6

FRONTIERS OF RESISTANCE IN THE AUTOMATED SOCIETY

The Street finds its own uses for things— uses the manufacturers never imagined.
The microcassette recorder, originally intended for on-the-jump executive dictation,
becomes the revolutionary medium of magnitizdat,
allowing the covert spread of suppressed political speeches in Poland and China.
The beeper and the cellular telephone become tools
in an increasingly competitive market in illicit drugs.
Other technological artifacts unexpectedly become means of communication,
either through opportunity or necessity.
The aerosol can give birth to the urban graffiti matrix.
Soviet rockers press homemade flexi-discs out of used chest X rays.
—William Gibson[1]

INTRODUCTION

How will we live in 1,000 years, if we are going to be still on this planet? Will artificial intelligence (AI) have completely enslaved us, or will it have freed us from laboring at all? A future with machines that are more intelligent than humans is not science fiction, but something that scientists predict will happen very soon. Stuart Russell, the founder of the Center for Human-Compatible Artificial Intelligence at the University of

California, Berkeley, said most experts believed that machines more intelligent than humans would be developed this century: "I think numbers range from 10 years for the most optimistic to a few hundred years," said Russell. "But almost all AI researchers would say it's going to happen in this century."[2] Will a global elite rule the rest of the population through AI, making our free will and agency obsolete?

Critics of the automation of society tell us that this danger is already here. The processes of automation enabled by AI systems are shaping human labor, consumption choices, political decisions, urban mobility, health, and cultural tastes.

The French philosopher Bernard Stiegler argued that our daily lives are completely overdetermined by automation.[3] According to him, we social beings subjected to digital capitalism are now reduced to a condition of automatons overdetermined by algorithmic mechanisms that nudge and shape our social behavior. Shoshana Zuboff takes a similar stance to that of Stiegler when she argues that for surveillance capitalism, "it is no longer enough to automate information flows about us; the goal now is to automate our behavior."[4] In their interpretation of contemporary capitalism, we seem to hear an echo of Horkheimer and Adorno's critique of instrumental reason: "Thought is reified as an autonomous, automatic process, aping the machine it has itself produced, so that it can finally be replaced by the machine."[5]

Recently, the media scholar Mark Andrejevic also expressed his concern about the rise of the automated society. He does not speak of online platforms, but rather of "automated media."[6] For him, when an algorithm decides what news, music, or video to show us, we are facing the automation of culture. What he calls "automated media" are beginning to permeate the world around us, mediating our symbolic exchanges with others. The digital communication networks on which the gig economy relies on are able, Andrejevic argues, to "subsume" all those forms of work that were once not subject to supervision and control. He convincingly argues that "the physical enclosure of the factory is no longer needed when virtual enclosures created by digital communication networks capture more information than any human supervisor could. The gig economy is a creation of highly automated systems for coordinating the activities of workers dispersed across the landscape."[7]

Platform capitalism seems to be an updated version of the *communicative capitalism* foreseen by the political theorist Jodi Dean more than ten years ago.[8] Following Dean's reasoning, today, we could say that not only has communicative capitalism appropriated networked telecommunications, transforming commercial digital platforms into the new infrastructures of society, but it increasingly relies on complex algorithmic infrastructures to exercise its power. The 2.0 version of communicative capitalism taking shape in this third decade of the new millennium is thus not only the consequence of the "convergence of networked communications and globalized neoliberalism,"[9] but also of the merging of globalized neoliberal ethos and datafied algorithmic infrastructures. This version is even more powerful than the previous one because it holds more data and computational power than the one originating at the beginning of the new century. The communicative capitalism of the twenty-first century extended its control over all forms of work that happen outside the traditional places of production: gig economy apps monitor, supervise, and make productive the effort of a bike rider, the driving skills of an Uber driver, the emotional labor of an elderly caregiver, the performance of an Upwork freelancer, and so on . . .

Work automation is not simply about replacing workers with an AI system, but it is a more complex process, tending both to optimize human work through AI and to optimize AI through human work. In the first case, we mean all those human jobs that are governed by algorithms with the aim of making them more efficient, as is the case of Amazon's workers[10] or those in the gig economy explored in chapter 3. In the second case, human labor is employed to make machines work better. The Italian sociologist Antonio Casilli has examined the forms of human labor hidden behind the training processes of AI systems,[11] while the Latin American sociologists Milagros Miceli and Julian Posada contributed to define *data work* as "the human labor necessary for data production . . . for machine learning." Data work, they argue, "involves the collection, curation, classification, labeling, and verification of data."[12] These workers who are literally behind the machines reside mainly in the Global South. Contemporary climate and geopolitical crises provide the gig economy and the artificial intelligence (AI) industry with an ever-growing reserve army of labor. The critical thinker Phil Jones[13] reveals how a globally dispersed complex of refugees, slum dwellers, and casualties of occupations are "compelled through immiseration, or else law, to power the machine learning of companies like Google, Facebook, and Amazon."[14]

Jones cites the example of the transport automation industry. The growth of the self-driving vehicle sector, he argues, depends on the ability of AI algorithms to correctly recognize all elements of the urban environment, from pedestrians and animals to traffic signs, traffic lights, and other vehicles. The images taken by in-vehicle cameras contain large amounts of raw visual data that first must be categorized and labeled in order to be fruitful. These data are used to train the software installed in driverless cars and prevent them from mistaking a traffic light for a pedestrian. Companies such as Tesla outsource data training to the Global South. In 2018, Jones reveals, more than 75 percent of this data was tagged by Venezuelans facing the most desperate circumstances. In the aftermath of the country's economic collapse, a significant number of newly unemployed people—including many former middle-class professionals—have turned to microjob platforms such as Hive, Scale, and Mighty AI (acquired by Uber in 2019) to annotate images of urban environments, often for less than $1 per hour.[15] More recently, *Time* magazine discovered that OpenAI used Kenyan workers (working for less than $2 per hour) to make Chat-GPT less toxic.[16]

Automation is advancing, no doubt. Even if the predictions of AI critics turn out to be wrong and AI does not outsmart us in a few years, scholars like Kate Crawford have demonstrated how the limited intelligence of current AI systems are already capable of putting our environment, freedom, democracy, and autonomy at risk.[17] She has shown how AI systems represent expressions of power of specific economic and political forces, which are created to increase profits and centralize control in the hands of those who hold them. The new media scholar Nick Dyer-Witheford and his colleagues argue that on its current trajectory, AI represents an ultimate weapon for capital. It will render humanity obsolete or turn it into a species of transhumans working for a wage until the heat death of the universe.[18] If the computational power of AI is going to increase, we need to ask ourselves in favor of whom this power will work and which actors will benefit from it. The risk that AI, whether really "intelligent" or not, may limit rather than enrich our lives and autonomy is real—here and now, not in 100 or 1,000 years. AI may indeed be harmful, although it probably won't be in the sense intended by the likes of Elon Musk and Geoffrey Hinton, the so-called godfather of AI, both of whom fear that AI is in danger of taking over humanity. These visions risk overshadowing much more concrete risks than the mass extinction of humanity at the hands of AI, such as the increased power of employers over employees, the use of AI against marginalized communities, and its devastating environmental impacts.

Yet is this description of the consequences of automation of society "thick" enough? Are we really being automated by the "smart" machines of digital capitalism without resistance? Are we facing (again) the return of an apocalyptic and deterministic vision of media and technology? This new emphasis on how digital platforms, AI systems, and algorithmic or automated media influence politics, enslave human labor, and shape cultural industries, exerting an increasingly ubiquitous form of power, risks losing sight of the space still available to people. In all these grand narratives of the current condition of increasingly mediatized capitalist democracies, there is something apparently minuscule and mundane that escapes the broad meshes of communicative or platform capitalism' theories. What escapes their gaze are the thousands of microscopic acts of resistance to the narrative force of platform capitalism, which materialize in the appropriation, by ordinary people, workers, and activists, of the tools of networked communication first, and algorithms today, to put them in the service not of a neoliberal ethos but of a cooperative and solidarity agenda.

Cinematic imaginary offers a highly powerful tool to answer these questions and imagine the future to come. Thousands of writers, authors, directors, and filmmakers imagine the humanity to come, influenced by the political and social problems of the historical moment in which they live.

Among the many possible fictional tales of the future, there are two movies that we would like to evoke. One is the remake of *Dune*, directed by Denis Villeneuve. In the first installment of this film series released in 2021, the cruel Harkonnen rule the planet Arrakis, also known as "Dune," on behalf of the Imperium, the master of the universe. The planet's native population, the Fremen (who look very much like the Polisario Front fighters of Western Sahara), are hiding in the sandy desert. The technological gap between the masters of Dune and the Fremen is enormous, but the Fremen have learned to live in the desert and are ready to resist. The dialectic between ruler and ruled is also at the heart of another science fiction film, *Elysium* (2013), directed by Neill Blomkamp. In the film, set in the middle of the twenty-second century, the Earth is overpopulated and heavily polluted. Most of the inhabitants live in poverty, subjected to extreme forms of surveillance. The rich and powerful have left and rule the Earth from Elysium, a space station orbiting the planet and equipped with advanced technologies to cure any disease. In this postapocalyptic scenario, a group of cyberpirates manage to build a spaceship capable of hacking into Elysium's defense systems and illegally land on it to use Elysium healing capsules to cure their friends.

What do these films tell us? They don't really speak about the future, but they narrate a thousand-year-old story. Where there is power, there is always a part of humanity that rises up to resist it. This is the case even when this power seems invincible, like that of the Imperium in *Dune*, of the rich citizens of *Elysium*, or of the Metaverse of Mark Zuckerberg. It has happened in the past; it happens every day; and it will continue to happen on Arkaris, on the satellite Elysium, or in 100 or 1,000 years on planet Earth (that is, if we survive ourselves).

Although we are immersed in this deeply "mediatized"[19] ecosystem and we live a "media life,"[20] our relationship with digital platforms is never frictionless. Users and platforms, human and nonhuman subjects coinhabit this ecosystem in complex ways. The relationship that we have with the algorithms that populate our daily lives is symbiotic and recursive. While it is true that algorithms learn quickly from users' gaming attempts and therefore are able to realign themselves, users are also capable of readjusting themselves to face the new challenges posed by algorithms, as demonstrated by the rider from Naples who found a way to circumvent the facial recognition software introduced by the food delivery company Glovo, as discussed in chapter 3. Every day, people temporarily ally themselves with algorithms to achieve their own strategic ends, but they also constantly break these alliances. Alliances *with* algorithms and rebellions *against* or *through* them alternate incessantly in everyday life.

This continuous realignment of algorithmic alliances gives life to contingent reconfigurations of power balances. We recognize that this relationship is informed by strongly asymmetric power relationships that leave increasingly less space for agency. However, we argue that algorithmic systems are to be considered as sociocultural and political battlegrounds where power is continuously renegotiated. People can still tactically appropriate the algorithmic infrastructure of platform capitalism to repurpose it at their own advantage.

In this last chapter, we summarize the key contributions of the book and reflect on its conceptual journey. We will reflect on the relevance of the key concepts of algorithmic agency, moral economy and resistance. We hope that, at the end, it will become evident that algorithmic agency, is an endemic and defining characteristic of our contemporary highly mediatized lives.

THE RELEVANCE OF ALGORITHMIC AGENCY

In this book, we have observed the emergence of various forms of algorithmic agency. We have done so by looking at the fields of gig working, cultural industries, and political activism for examples. We could have focused on only one of these fields—in fact, each of them could be a book in itself—but we chose to bring them together in this book because we wanted to show how forms of algorithmic agency are a structural condition of the platform society, not just a set of practices related to one or more platforms. The platform society described in detail by José van Dijck et al.[21] and the whole emerging platform studies field[22] is defined by the semimonopolistic and quasi-infrastructural power that platforms are acquiring, but, at the same time, platform power is only half this story. Just as there is no power without resistance, there is no platform power without algorithmic agency. Through case studies taken from the worlds of gig working, cultural industries, and politics, we wanted to show how, wherever there is an algorithm acting as an intermediary, there is always someone who has found a way to repurpose that algorithm to their own advantage. This does not mean that these practices are always "right" and "fair" or that they are widely spread among all people. On the contrary, we have seen how knowledge of these practices is not equally distributed among users and how the effectiveness of these practices depends on the availability of time, capital, and expertise.

Yet the mere fact that they exist and continue to evolve demonstrates the human capacity to respond both tactically and strategically to the challenges posed by algorithmic power. Algorithm gaming practices are everywhere: in work, cultural production and consumption, politics, and many other fields affected by platformization. Recently, the scholars Krishnan Vasudevan and Ngai Keung Chan discovered how Uber drivers resist Uber's gamification features by using them in unintended ways. Drivers appropriated the affordances of the Uber app and reoriented its purpose to maximize their earnings and maintain control over their labor.[23]

We showed that in these three realms of everyday life, the repertoires of algorithmic agency can follow similar patterns, even if they serve different aims. But there are many more practices to be discovered beyond these three domains. Our attempt only touches the tip of the iceberg. If we look for them, practices of algorithmic agency can be found in many other spheres of everyday life, even the most mundane and banal.

Clever tactics against the power of algorithms over our daily lives are flourishing everywhere. For instance, a Maryland resident, Timothy Connor, saw an exponential increase in traffic spring up overnight on the quiet street where he lived. When he realized that this increase was due to the suggestions provided by Waze to drivers who wanted to avoid a road under repair in the vicinity, Connor borrowed a tactic that he read about (on an online private group) from the car wars of southern California and other traffic-weary regions. He became a Waze "impostor." Every rush hour, he went on the Google-owned social media app and posted false reports of a wreck, speed trap, or other blockage on his street, hoping to deflect some of the car traffic.[24]

We could also look at the example of the German artist Simon Weckert, who borrowed phones from friends and rental companies until he had acquired ninety-nine devices, which he piled into a little red wagon. Weckert started taking the red little wagon along the Spree River, in Berlin, creating a huge traffic jam—one that only existed on Google Maps. With his "Google Maps Hack," Weckert wanted to draw attention to the systems that we take for granted and how we let them shape us.[25] But we could also cite the case of the *Vice* journalist Oobah Butler, who managed to turn a shed in his garden into London's top-rated restaurant on Tripadvisor,[26] demonstrating how easy it is to fool reputational algorithms.

All these cases go beyond the world of gig work, culture, and politics. Artists, students, and other citizens who want to preserve the silence of their neighborhood, K-pop fans, criminals who want to make easy money, food delivery couriers, politicians, activists, high school students, social media content creators, hoteliers and restaurateurs, and who knows how many others encounter different kinds of algorithmic media every day and constantly negotiate with them a truce or an alliance, leveraging all the means at their disposal. Sometimes the truce breaks down, and the algorithms become an enemy; sometimes there is no negotiation at all and the algorithms are naturalized and incorporated frictionlessly into our daily lives; or sometimes we strategically ally with them. The tactics and strategies that we have shown are relevant not only for the people who enact them to survive and cope with algorithmic power. They are relevant for the rest of us too because they represent the evidence that it is possible to negotiate our relationship with algorithms, even within the cramped limits imposed by platforms. Yet we have only scratched the surface of algorithmic agency; there is still a lot of digging and much research to do.

ALGORITHMIC AGENCY ACROSS GIG WORK, CULTURE, AND POLITICS

There is a common characteristic to all the practices we have observed. Algorithmic agency is rarely exercised in complete solitude, carried out by a single individual without the help or partial cooperation of some other person. Many of the forms of algorithmic agency that we have been examining may be individual actions, but this is not to say that they are uncoordinated, as James Scott already explained about the forms of everyday peasant resistance that he observed in Malaysia.[27] Even in those cases where individuals acted alone, the tactics used were learned from other individuals they met in online groups on WhatsApp, Telegram, or Facebook. Only by belonging to one of these groups could individuals learn tactics that they enact on their own. We have seen how private online chat groups represent infrastructures of solidarity and learning environments for couriers of the food delivery online services, but also for social media independent content creators. Most of the time, gaming algorithms is a collective effort. Especially when the gaming attempt comes "from below" and does not have large economic and computational resources available, the cooperation of a wide network of users is central to its success. In many cases, the exercise of tactical algorithmic agency gets results only if it can involve an entire swarm of users who volunteer their time to make up for their lack of computational power and economic resources.

In the previous chapters, we have seen how couriers, content creators, and social movement activists have joined forces to "grow together" or give more visibility to their cause. In gig working, cultural industries, and politics, people dealing with algorithms and platforms realized what the anarchist thinker Pyotr Alekseevič Kropotkin[28] had already understood more than 100 years ago: that cooperation brings them more benefits than competition. Collective action and cooperative ethics are thus common features of all the fields we studied, but there are other similarities, as well as some differences. In the field of cultural industries and political activism, for instance, the practices that we showed in our discussion all had to do with visibility: content producers and consumers, political activists, marketing agencies, and political institutions exercised strategic and algorithmic agency with the aim of gaining more visibility (or, in some cases, to avoid visibility). Platforms like Facebook, Twitter, Spotify, TikTok, Instagram, and YouTube exercise capillary control over the visibility of every item shared on them. This control over the visibility of objects and people, made possible by the computational power of the platforms, is the target of all four manifestations of algorithmic agency described here. Visibility, in these platforms, is a contested commodity. Strategic and tactical manifestations of algorithmic agency, whether aligned with the moral economy of the platforms or not, are aimed at partially or fully contesting the algorithmic governance of visibility exerted by platforms.

While in the case of politics and cultural industries, platforms exert algorithmic governance of visibility, in the gig economy, algorithmic governance manages the physical and cognitive performances of workers.[29] Overall, then, we can see how, on the one hand, platforms employ algorithms to better control the bodies of workers and the visibility of the products of human creativity. On the other hand, gig workers, cultural prosumers, and political actors try to resist this control over their bodies and visibility by exercising their algorithmic agency. Another feature that cuts across all the manifestations of algorithmic agency is the existence of a moral code that can be aligned with that of the platforms, or not.

"OMAR COMIN'!": ON THE RELEVANCE OF THE MORAL ECONOMY CONCEPT

One of the most significant scenes in the beloved HBO television series *The Wire* happens during the second season. Omar Little, who goes around robbing drug dealers, is testifying in court against a drug dealer named Bird, who is accused of murder. Bird's lawyer, Maurice Levy, is paid by one of the most powerful drug dealers in Baltimore, Avon Barksdale, to get his "soldiers" out of trouble. Levy, in Bird's defense, accuses Omar of being a less than credible witness. "You are amoral, are you not?" he asks him. "You are trading off the violence and despair of the drug trade. You are a parasite that leeches off . . . ,"

"Just like you, man," interrupts Omar.

"Excuse me, what?" asks a shocked Levy.

"I got the shotgun. You got the briefcase," Omar replies. "It's all in the game, though, right?"

At first sight, Omar is telling Levy that there is no moral difference between the two of them—both are corrupt. If, as Levy says, Omar is not credible, then neither is Levy. But there is more than that. After the death of Michael Kenneth Williams, who played the epic character of Omar Little, the *Guardian* journalist Kenan Malik argued that if we look at the entire narrative arc of *The Wire*, it should be clear to us that the underlying message of this crime series is that people who find themselves in impossible situations are forced to figure out for themselves what is rational and moral within their own life-world: "What may seem from the outside, from those who make the rules of 'the game,' as irrational and immoral is, for those trapped by the system, the only way to weigh up good and bad in the settings in which they find themselves."[30] Malik invites us to look at the world of *The Wire* as a complex ecosystem, where different moral economies collide and the definition of what is right and wrong is always a question of power.

This book takes this invitation seriously and projects it onto the platform society. What the platforms consider as immoral or illegitimate might instead be the only possible behavior for those on the other side or for those who use them to work or add visibility to their political cause. There is no one single dominant moral economy within the platform society, but several competing moral economies, and the boundaries between these economies are blurred. For both users and platforms, the "ends justify the means," but the ends and means of the former do not always coincide with those of the latter.

Platforms' moral economies are characterized by a blind faith in their ability to determine what and who *deserves* the most attention. Western commercial platforms embody the neoliberal ideology of meritocracy. The software developers of these platforms genuinely believe that they are able, through their data and algorithms, to recommend what is best for users[31] or to calculate who are the best-performing workers. The moral economy of the platforms is characterized by the fetishism of algorithmic rankings. Rankings always have a winner and a loser, whether it is a piece of music that will be recommended more than others, a rider that will receive more orders than others, or an Instagram post, a TikTok video, or a tweet that will be granted more visibility than others. Users who embrace this moral belief find this competition "fair" and "natural."

However, we have seen that there are other users who do not accept the logic of the sheer competition and prefer to ally with their peers against the platform (or other users). Those who do so are usually aware that platforms do not simply reward the "best" item. Couriers starting out in a city where there are already hundreds of other delivery workers know that it is impossible to gain reputation points and climb the ranking without someone's help or a few tricks. Likewise, Instagram content creators know that those who have money to promote their content get a visibility boost that is not at all meritocratic.

Political activists know that Twitter does not treat all user-generated hashtags in the same way.[32] Often, those who resort to gaming practices do so because they have no alternative and could not survive on the platform otherwise. But just as often, those who resort to these practices do so because they do not fully accept the moral discourse encoded by the platforms and recognize social media as highly toxic, extractive commercial environments.

A typical feature of the users' moral economies is the rhetoric of "growing together," of helping each other. These users oppose the mutualistic discourse of "united we are stronger" to the meritocratic discourse of the platforms. This mutualism can be put at the service of different ends, which we—from our specific sociomaterial point of view—may consider dangerous, right, wrong, immoral, banal, or heroic. We did not want to argue in favor of any moral economy of users. Building on the invitation made by the US anthropologist Nick Seaver, to consider algorithms as cultural artifacts,[33] this book has illuminated the complexity of the platform society and the existence of manifold ways of making sense of algorithms.

The concept of moral economy helped us to foreground this complexity and to illustrate how the supposed objectivity and neutrality of algorithms is not a "natural" property of algorithms but, on the contrary, is just a matter of power: the power of those who decide what and how should be calculated and for what purposes. Algorithms perform operations based on a set of instructions (if . . . then) set by someone driven by specific objectives. If the objectives change, so do the results and thus the ranking of items generated by these algorithms. Algorithmic computation can therefore be altered, modified, manipulated in many ways, at different levels, by both human and nonhuman actors. In the case where the modification occurs "from above," from the platform itself, it is considered legitimate. Platforms constantly modify their algorithms, and consequently their ranking systems. When Instagram and YouTube changed their algorithms, many talented content creators suddenly lost visibility. There is nothing objective or meritocratic about this. However, when it is the users who manipulate the results of the algorithms, "from below," this practice is considered illegitimate by platforms. As in the world of *The Wire*, what is fair and what is not? Which manipulation is more "just"? It is always the more powerful actors who decide what a technological artifact should do. The designers of objects, technologies, and artifacts inscribe in them an intention, give them a meaning and a "moral": they design these objects in a particular way, so that they can afford some actions at the expense of others.

Designers know what their objects *should* do. So, in the heads of a Spotify or Netflix software designer, an algorithm *should* give the listener something that will hold their attention for as long as possible, while Frank, Deliveroo's algorithm, *should* be able to assign an order to the "best" rider available at that very moment. In the case of Spotify or Netflix, the designer does not inscribe in the code of the algorithm the task of recommending to the user items that represent the diversity of the musical or audiovisual cultural production of a particular society, as the algorithm of a public service platform, such as the BBC, might do. Thus, the Deliveroo algorithm has not inscribed in itself the task of distributing to all couriers enough orders to make a decent wage. This task could instead be inscribed in the algorithm of a cooperative food delivery platform.

Algorithms, like any artifact, communicate and automate the moral values of those who designed them through their affordances. All artifacts, in fact, as noted by the Australian sociologist Judy Wajcman, "reflect the culture of their makers."[34] Building on the American social scientist Langdon Winner, we could say that artifacts do not only have politics,[35] but they also have a *moral*. Affordances mediate the specific moral values inscribed in them. But does this mean that the way that technologies are designed determines user behavior? Not quite. Winner's view should be complemented with a reflection on the agency of users of artifacts, which we inherit from British cultural studies thinking on media audiences[36]. Moral values inscribed in technological artifacts are neither hegemonic nor fixed forever. These moral discourses can be accepted, negotiated, or rejected by users. When users "decode" algorithms,[37] they also decode the meanings and moral values attributed to them by their creators. Users can accept the moral economy inscribed in the apps they use or else challenge it, even if only partially or temporarily.

In the examples that we have seen so far, users attribute new meanings to the algorithms that they use and attempt to put them at the service of different moral values. The algorithms of Instagram, Twitter, Deliveroo, and Spotify have been designed according to a morality that rewards competition among individuals. On the other hand, some of the users of these platforms, like the Indonesian couriers we met in chapter 3, try to bend the affordances of these algorithms according to a morality that rewards a collective, a mutually supportive group, sometimes to the detriment of other groups or collectives of actors. The concept of moral economy has been productive to frame the relationship between platform power and user agency as a battleground where manifold moral values face each other. But again, we feel we have only scratched the surface. There is still a lot of research to be done. Our hope is to have breathed new life into this concept.

THE RELEVANCE OF EVERYDAY ALGORITHMIC RESISTANCE

Many, though not all, of the manifestations of algorithmic agency in the domains of work, culture, and politics that we have described so far can also be understood as forms of resistance. Drawing from the work of Hollander and Einwohner[38] and other resistance scholars, we clarified in chapter 1 what we mean by resistance: (1) an act, (2) performed by someone upholding a subaltern position or someone acting on behalf of and/ or in solidarity with someone in a subaltern position and (3) most often responding to power *through algorithmic tactics and devices*. We focused our attention on the kind of resistance exerted *through* algorithms rather than *against* them. We called this form of resistance algorithm-*enabled* resistance, which conceptualizes algorithms as a *repertoire* of tools and tactics. In chapter 1, we also argued that there is neither a clear distinction between agency and resistance nor a perfect superposition. Rather, we proposed that the agency manifestations that we described there move along a continuum that goes from forms of agency openly resisting platform power and other forms of agency that have no intention of questioning or challenging platform power and thus cannot be deemed as resistance acts.

All those practices that characterize the moral economy of users as opposed to that of platforms are certainly resistential. For example, all collective actions oriented toward user cooperation are in fact, regardless of their aims, a form of resistance to the hegemony of the moral economy of platforms, based on individual success. The kind of resistance we have shown in the chapters on labor, culture, and politics is a kind of mundane resistance, an "everyday" resistance, as James Scott understood it, or an "art of the weak," as Michel de Certeau framed it. Scott noted that "the goal, after all, of the great bulk of peasant resistance is not directly to

overthrow or transform a system of domination but rather to survive—
today, this week, this season—within it."[39] We could say the same for
most of the forms of resistance mapped so far: their goal is to survive
today, this week, this season.

To make it even clearer, we could consider couriers of food delivery apps,
K-pop fandoms, Instagram podders, and political activists as bricoleurs
who are able to adapt, manipulate, and renegotiate tools (the algorithms)
that they neither possess nor control directly. From this point of view,
the acts of resistance that we have described resemble the ritual forms of
resistance enacted by the British subcultures described by Stuart Hall and
his colleagues in *Resistance through Rituals.*[40] As James Procter pointed out
in his book on Hall's intellectual legacy, "Unlike revolutionary resistance,
which tends to work by rejecting or overturning, ritual resistance is about
using and adapting. Such forms of resistance are not necessarily going to
'revolutionize' class structures in the sense of a straightforward inversion;
they are *potential* forms, 'not given but made.'"[41]

Like *ritual* resistance, algorithmic resistance remains a process of ongoing negotiation rather than a solution to the power of platforms. Hall and
his colleagues argued that the styles and rituals of subcultures could only
be used to negotiate and survive the experience of belonging to a subordinate class; they could not solve it: "The problematic of a subordinate class
experience can be 'lived through,' negotiated or resisted; but it cannot be
resolved at that level or by those means."[42]

Similarly, the repertoire of algorithmic resistance tactics that we have
shown are not a solution to the overwhelming power of platforms, but
rather a proof of a mismatch between users' and platforms' moral economies. Although endowed with enormous computational and symbolic
power, the hegemony of platforms can hardly be taken for granted. Their
power is not unending nor is the exercise of this power frictionless, just
as platform users are not permanently incorporated and subordinated to
the moral economy of platforms.

In the chapters on gig work, culture, and politics, we described many
practices that we interpreted as expressions of algorithmic agency. As a
final synthesis of our work, we positioned all the tactics and strategies
mentioned so far within the conceptual framework described in chapter 2. Figure 6.1 shows how workers, cultural prosumers, and political

6.1 Practices of algorithmic agency and resistance situated in our conceptual framework.

activists exercise their agency either strategically or tactically, with different endowments of capital, time, and expertise, and along competing moral economies.

CONCLUSION: THE MAKING OF THE PLATFORM WORKING CLASS

In his famous book on the making of the English working class, the historian Edward P. Thompson showed how rural peasants and inner-city craftsmen were industrialized between the eighteenth and nineteenth centuries.[43] The peasants gradually left the countryside, whose lands had been enclosed and privatized, to be employed in the new cycle of industrial production. Craftsmen who were used to working in small workshops or in their own homes gradually abandoned their workplaces and moved into factories. By the latter part of the nineteenth century, the process of industrialization of labor in the UK was mature.

In a similar way, we could say that today, a whole range of activities that used to be carried out online on the web, or offline in the world of "atoms,"[44] are increasingly platformized: bloggers, content creators, political activists, delivery couriers, taxi drivers, musicians, and consumers of cultural products, news, or food are undergoing a process of progressive platformization. Work, cultural production and consumption, and political activities have been progressively captured and enclosed in the private spaces of commercial platforms. Bloggers who had created their own independent websites had to converge on Instagram, Facebook, Twitter, and YouTube to continue to meet an audience, while independent podcasters who used to have their podcasts delivered through an .rss feed are now moving into the walled garden of Spotify.

Yet platformization is neither a uniform nor a frictionless process. Thompson showed that early industrial capitalism took time to turn the masses of English peasants and artisans into disciplined workers, respectful of factory rules and working time. Industrialization has not advanced in a linear fashion; it has been shaped by the emergence of mutual aid associations, benefit societies, clubs, and eventually trade unions that helped to transform the "18th century mob in the 19th century working class."[45]

If we situate platform capitalism in the *long durée* of industrial capitalism and look at it as the latest cycle of capitalist accumulation of value, we may be facing a transition similar to the industrial one: labor, consumption, and political activities are gradually subsumed and captured inside the fences of platforms, but this process of platformization, like the previous one of industrialization, is again neither homogeneous nor linear. As Nieborg et al. have argued, we should challenge any "essentialist theory of platform dominance."[46]

On the one hand, the mass adoption of platforms for work, culture, and political activities also entails the adoption by users of the moral economy of platforms, based on individualistic and highly competitive behavior. On the other hand, this moral economy does not impose itself on everyone in the same way. Platform capitalism is taking time to transform its users—whether Deliveroo couriers or Instagram prosumers—into disciplined workers, competing for money and visibility. While this process of platformization is gaining momentum, at the same time, practices emerge that either are holding back or have the potential to shape this process. Just as the English working class emerged between the eighteenth and nineteenth centuries, a new platform "working class" with its own specific culture in dynamic contrast to that of the platforms is slowly emerging.

In the case of the Industrial Revolution, Thompson noted that by the nineteenth century, factory workers had begun to organize themselves, giving rise to benefit societies governed by extremely strict moral codes and based on mutual aid:

By the early years of the 19th century it is possible to say that collectivist values are dominant in many industrial communities; there is a definite moral code, with sanctions against the blackleg, the "tools" of the employer or the unneighbourly, and with an intolerance towards the eccentric or individualist. Collectivist values are consciously held and are propagated in political theory, institutions, discipline, and community values which distinguishes the 19th century *working class* from the 18th century *mob*.[47]

In the case of the platform society, our examples show that platform users also started to organize themselves, to come together first in informal solidarity groups, such as private online groups on WhatsApp, Telegram, and Facebook, and then in more formal associations. In Naples, members of a private WhatsApp group of couriers set up Casa del Rider, a common space where they could meet and take a break from work, as we have seen in chapter 3. Something similar, but much more structured, is happening in Jakarta. The researchers Rida Qadri and Noopur Raval have mapped the emergence of hundreds of informal courier associations in the Indonesian capital. These associations have built roadside shelters where couriers and drivers can meet, exchange information, recharge their smartphones, and wait for their next order. Qadri and Raval detail the daily life of an Indonesian driver, Mba Mar, and that of her community:

Mar's community is just one of the hundreds of platform driver collectives spread across Jakarta. Each has its own membership rules, ranging from moral expectations (members must be honest) to socializing expectations (members must remain an "active" part of the WhatsApp groups, attend all social events of the community, come to the basecamp at least once a week, and so on). Communities hold internal elections and have mandatory monthly member meetings. Some even have membership fees, which go into a common pool of money used to support community expenses. Most communities have built basecamps where drivers meet between orders, some calling these spaces their "second home." Many issue ID cards to identify members in case of road accidents, and as a way to solidify their sense of belonging. Collectively, they have set up their own joint emergency response services, and informal insurance-like systems that use community savings to guarantee members small amounts of money in the case of accidents or deaths. They have also provided their members with Covid relief, such as distributing personal protective equipment and free groceries.[48]

These Indonesian communities of *ojol* (mobility platform drivers) are interconnected through WhatsApp groups in which they build networks of solidarity and friendship: "*Salam Satu Aspal*, the motto of the *ojol*, signals their unity: they share with each other the 'Blessings of One Road.'"[49]

In addition to Indonesia, similar mutual aid associations coordinated via WhatsApp emerged in China and Mexico, as we saw in chapter 3, and in who knows how many other countries in the Global South where no research has yet been done. The similarities between these informal solidarity groups and the early benefit societies described by Thompson are striking: membership rules and fees, moral expectations, and informal insurance-like systems. Here, too, collectivist values prevail over the individualist values typical of the moral economy of Western commercial platforms.

During the eighteenth and nineteenth centuries, benefit societies in the form of friendly societies and trade unions were essential in providing social assistance for sickness and unemployment and improving social conditions for a large part of the working population. Similarly, more contemporary forms of benefit societies are essential to improve the conditions of platform workers.

Some might argue that in contexts of extreme precarity such as that of food delivery couriers, forms of solidarity are more likely to emerge. Yet in chapter 4, we saw that such associations are also emerging in the global influencer marketing industry, albeit in different forms: alliances between traditional unions and content creators for YouTube, new unions in defense of content creators, or simpler engagement groups for Instagram and TikTok, which also have, as we have seen, their own moral code. K-pop fan communities also established their own moral code, which may be at odds with that of the platforms, as do political activist movements that use platforms to amplify their ideas. As we extensively illustrate in chapter 5, social and political movement activists often frame their appropriation of the algorithms of corporate social media to amplify their voices as legitimate action because they recognize the toxic and extractive nature of these platforms. Thus, any kind of practice that can "use them back" to fuel activists' causes is seen as legitimate. In other cases, this led activists to completely withdraw and abstain from using digital platforms because they are perceived as actors whose logic clashes with the development of proper civic cultures, dialogue, and democratic values.

The platform society is thus undergoing two opposing processes: on the one hand, the process of platformization of workers, consumers, activists, and more generally of all the customers of the platforms, which tend to discipline and socialize them to the rules and the habitus of their moral economy; on the other, a process of growing resistance to this moral economy, which is taking on increasingly structured forms of individual and collective action. But these increasingly structured forms of resistance can emerge only when platform users have acquired a certain degree of awareness of the processes of spoliation of their data, bodies, emotional states, and performances operated by platforms.

In this sense, the practices of everyday resistance that we have mapped represent the first step, not the solution, in a process of awareness-raising among a vast multitude of actors who occupy a position of moderate/ extreme weakness or subalternity with regard to platforms. As James Scott noted, "In the process of struggling, they discover themselves as classes."[50]

Commenting on the Luddist movement, the British historian Eric Hobsbawm once suggested that "through machine breaking itself, the luddites composed themselves as a class by creating bonds of solidarity."[51] Drawing on Hobsbawm, we could argue that through the exercise of tactical and strategic forms of algorithmic agency not aligned to the platform moral economy, platform workers "composed themselves as a class by creating bonds of solidarity."[52] Through the daily gossip about the working of algorithms, the exchange of tricks and the collective exploitation of platform loopholes, platform users learn the benefits of mutual aid and cooperation and build networks of solidarity (either "entrepreneurial" or "oppositional"). Each new cycle of industrialization has not only forged a new type of worker but has in turn been shaped by the various forms of resistance opposed by workers. While Thompson and Hobsbawm studied the emergence of the English working class and the construction of solidarity bonds among industrial workers, Italian theorists of workerism[53] accounted for the new process of class composition and the new forms of resistance exhibited by the youthful mass workers deskilled by the introduction of new automation technologies into the Italian factories of the 1960s and 1970s.[54] Today, we may be facing a new phase of industrial capitalism, in front of which a new class—where, by "class," we do not mean a specific social formation in the Marxian sense, but rather a more heterogeneous global "multitude"[55] of platform workers, cultural prosumers, and political activists—is being recomposed.

We mentioned in the introduction that the digital labor performed by the members of this multitude is of different types (free labor vs. platform labor), the members are subjected to different forms of exploitation, and manifold are the kinds of agency afforded to each. The goal of this book, however, has been primarily to show what these categories of digital laborers have in common—namely, the ability to painstakingly improve their working conditions, organize forms of collective action, and build solidarity bonds in the face of the disproportionate computational power wielded by tech companies and regardless of their situated class status.

Among this multitude, whose daily life is continuously datafied and exploited through a dispositive that Shoshana Zuboff calls "behavioral surplus accumulation,"[56] many just stop at the first step of everyday resistance because their energies are barely enough to survive in this highly surveilled ecosystem. Criticizing the power assumed by these platforms, as many radical scholars do, is a necessary move because its excesses are there for all to see. Theorizing how digital capitalism could be revolutionized is even better and certainly very fascinating. But to change the face of digital capitalism these critiques are not enough. Change will happen only when people who use these platforms every day to work, consume, inform themselves, and engage in political activity realize that they are being exploited or at least are severely limited in their agency. Faced with this realization, the first reactions are to try to cheat the system, getting as much benefit as possible without any intention of really changing it. Many of the practices described in this book stop at this point. But other people, especially those who depend economically on the platforms for a living, realize that this ecosystem makes their lives precarious and unpredictable and then begin to organize, protest, and eventually imagine alternative solutions. Only through the embodiment of the most pervasive forms of power is it possible for the spark to be ignited to try to change things.

Thousands of food delivery couriers, after meeting in private chats, started to organize unauthorized strikes, found new unions, or decide to definitively log out to found cooperative platforms, based on alternative moral codes to those of commercial platforms. In Bengaluru, the capital of the Indian state of Karnataka, for example, the Autorickshaw Drivers Union (ARDU) launched its own ridesharing smartphone app, Namma Yatri, on November 1, 2022, based on ethical values like fair wages and affordable fares.[57]

In addition to founding cooperative platforms that could foster more *convivial* social relationships,[58] this emerging platform working class is taking advantage of the digital technologies of commons-based peer production to create small-scale entrepreneurial activities, especially in the Global South.[59] Alongside the semimonopolistic global platforms, many small digital platforms of petty producers are flourishing, which the Swedish sociologist Adam Arvidsson interprets as a sign of the rise of a new industrious class, which refuses to work for Big Tech and is shaping the future of the digital economy.[60] In the field of cultural industries too, as we have seen, new unions of creative workers have emerged, while in the field of political activism, there is a rising mobilization against the discriminatory power of algorithms.[61]

Critics of the everyday forms of resistance claimed that such acts intrinsically lack revolutionary consequences. Yet, echoing Scott's words again, "this may often be the case, but it is also the case that there is hardly a modern revolution that can be successfully explained without reference to precisely such acts when they take place on a massive scale."[62] And even in cases where these resistance practices are limited to daily survival, we can say, as Scott does, that they "prevent the worst and promise something better."[63]

We have described the ambivalence of the transition taking place in the platform society, framing it as a contested process between two (and probably, more) competing moral economies. Platform users continue to swing between forms of agency shaped by cooperative moral values and others instead influenced by competitive moral economies.

While this book recognizes the power imbalance between platforms and users, it also casts a more complex and nuanced vision, showing that the conflict is still open: the sounds of battle are already here, but the outcome is still uncertain. Remixing Stuart Hall et al. about the relations between subordinates and dominant cultures, we could say that "the outcome" of the conflict between platform power and human agency "is not given but *made*."[64]

People have begun to realize that computational power in the hands of platforms can be countered only by joining forces and pooling our knowledge, time, and economic resources. Computational power can be confronted only through cooperation, mutual aid, and collective power.

Platforms are a battleground where people sometimes dance with algorithms and other times clash with them. Sometimes they lose; other times, they (temporarily) win. Sometimes they game the system; sometimes they radically change it.

The fight is still on.

APPENDIX: RESEARCH METHODS

This book is the result of a painstaking effort to weave together various strands of research and fieldwork. Most of the data generated for this book emerged during one of the most difficult periods of our lives, the global COVID-19 pandemic. We started discussing the topics of this book in April 2019, during a dinner in Siena, in front of a good bottle of wine. We both had behind us several bodies of research on people's agency with digital technology, although in different fields (Emiliano in politics and Tiziano in cultural industries), and some of the data that we have collected in the past was reanalyzed for this book. However, most of the interviews and fieldwork began close to the end of 2019, and a few months later the whole world was in lockdown: Emiliano in Cardiff, Tiziano in Florence. We had to wait until the end of August 2021 to meet in person again. On that occasion, in a farmhouse near Marradi, lost in the mountains on the border between Tuscany and Emilia Romagna, we finally discussed for ninety-six hours the concepts of the book, the data collected, and the division of the writing. But between March 2020 and August 2021, we had already built the foundations of the book through a daily dialogue on a WhatsApp chat and innumerable Zoom and Jitsi calls. To write this methodological note, we exported the chat and calculated the words that we exchanged from January 2020 to November 2021. Our conversations around this book corresponded to 139,897 words.

For almost two years, the chat was the place where each of us shared new ideas, reflections, interpretations of data, discoveries, and moments of disappointment. In fact, these 139,897 words represent the field notes of our research. This is the shape that the famous ethnographer's field notes have taken in the era of a global pandemic: a WhatsApp chat written by the two of us between Cardiff and Florence. While carrying out research during a pandemic was a great challenge at various levels, it also represented an opportunity to ignite our "methodological imagination."[1] It pushed us to find new ways to harness the power and flexibility of qualitative, hybrid, and multisited research in this unpredictable scenario. Further, it reinforced our commitment to doing qualitative research to excavate the meanings, challenges, and understandings that social actors ascribe to their practices in such challenging times.[2]

This book brings together several data sets gathered during different periods of time across multiple countries by scholars working at various institutions. If we were to situate chronologically the different pieces of research that constitute this book, the result would be as follows. Emiliano's research on digital activism and data politics from 2012 to 2020 across several projects in Latin America and various European countries (including Spain, Italy, Greece, and UK) fuels chapter 5. Research on Instagram pods from October 2019 to May 2020 by Tiziano and his research assistant, Francesca Murtula, informs chapter 4. Desk research carried out by Tiziano and Emiliano in 2020 and 2021 feeds into all the book's themes. During May and June 2020, Thomas Davis at Cardiff University extended the part of the database about algorithmic politics, thus contributing to chapter 5. The cross-cultural, multisited research on food delivery couriers' practices carried out between July 2020 and August 2021 informs chapter 3 and involves various researchers in a number of countries (including Mexico, India, China, Italy, and Spain) coordinated by Tiziano and Emiliano. We provide more detail regarding these data sets and the research that produced them in the next sections, where we follow the order of the book's chapters.

GAMING THE BOSS

The individual and collective tactics deployed by riders of online food delivery services have been mapped thanks to the Algorithmic Resistance

Project (AlgoRes), of which we were co–principal investigators (PIs). Its aims were to explore the algorithmic agency and resistance tactics of online food delivery riders in the Global South. The project started in July 2020 and ended in August 2021, with an online workshop accessible only to the research team, where we shared and commented on our findings. Four early career researchers participated in the project: Daniele Cargnelutti (University of Guanajuato, Mexico), Swati Singh (Indraprastha College for women, Delhi University), Zizheng Yu (University of Greenwich, UK), and Francisco Javier Lopez Ferrández (Jaume I University, Spain). All these researchers had previous connections to Emiliano's academic trajectory and were recruited organically for this project because they had experience and availability and could cover a wide variety of cultural contexts and countries.

Daniele had been Emiliano's research assistant at the University of Querétaro, Mexico. Swati joined this project as part of a collaboration with Emiliano as she was getting her bachelor of arts degree. Emiliano was Zizheng's PhD supervisor at the School of Journalism, Media, and Culture, Cardiff University, and he acted as Javier's PhD external examiner in Spain.

The research team carried out a total of sixty-eight interviews (seven in Mexico, thirty-two in India, twelve in China, twelve in Italy, and five in Spain) with online food delivery couriers and did participant observation of dozens of WhatsApp, WeChat, and Facebook groups with thousands of riders in all five countries involved in the study. Interviews were carried out in a number of cities: Querétaro and Mexico City in Mexico; Delhi, Gwalior, Mumbai, Pune, Lucknow, Chattisgarh, Gurugram, and Patna in India; Beijing, Shanghai, Shenyang, Weifang, and Dongguan in China; Livorno, Florence, Milan, Naples, and Messina in Italy; and Valencia, Barcelona, and Bilbao in Spain. The platforms involved were Uber, Cabiify, Didi Food, InDriver, EasyTaxi, Rappi, Sin Delantal, Didi Food, and Uber Eats in Mexico; Swiggy, Zomato, and Uber Eats in India; for Meituan, Ele. me, Flash EX (Shansong), and SF Express in China; Just Eat, Uber Eats, Deliveroo, and Glovo in Italy; and Uber Eats, Glovo, Deliveroo, Just Eat, and Stuart in Spain.

Researchers established a long-distance dialogue via WhatsApp with some of the interviewed couriers for several months, and this dialogue was deeply meaningful for understanding some aspects of their work.

Some of the interviewed couriers allowed us to follow them during their work shifts and showed us how food delivery apps work. In Mexico, Daniele Cargnelutti also worked for two weeks as a courier to get a firsthand impression of how the algorithm hides itself behind the high wall of obscure regulations and how gamification works in practice. Two couriers also agreed to read what Tiziano had written and provided their comments in chapter 3. The interviews lasted from thirty to sixty minutes and were recorded using digital audio recorders. Interviews were conducted in the original languages of the couriers (Chinese, Hindi, Italian, Spanish, and Mexican Spanish) and then translated into English. We are aware of the many concerns and limitations surrounding both the translation of concepts into another language[3] and the choice of English as the lingua franca of academia,[4] and for this reason, we invited all the members of the research team to publish individual papers on other aspects of this research in their home languages.

The interviews were complemented by a digital ethnography[5] of dozens of online private chat groups (mainly on WhatsApp, Telegram, Facebook, and WeChat) created by the couriers. For a full year (July 2020–August 2021), we observed thousands of couriers interacting with each other in these private chat rooms. Private chats are enabled by apps like WhatsApp and Telegram. They are cross-platform messaging apps that allow users to exchange messages over a phone's data traffic without paying extra for short-text messaging, and they have proved to be very effective in fostering peer-to-peer communication in both contemporary media activism[6] and gig labor.[7] Studying activists' and workers' social interactions within the chats enabled by these apps allows researchers to observe social dynamics over time and provides a huge wealth of data.

Conducting ethnographic research within instant messaging apps represents an innovative practice in the field of digital ethnography, but it also comes with significant ethical challenges. Barbosa and Milan ask themselves "how to develop a creative approach to digital ethnography that did not harm or interfere with the interactions among chat members."[8] In accessing this new field of research, we followed their approach, which is based on the "do no harm" principle in private chat groups.

Access to the field began in Italy. Stefano—the courier whose practices we analyzed at the beginning of the book—was the first gatekeeper who introduced us to the field. He was a friend of ours, and we had several conversations with him about his work as a courier. He introduced us to other couriers who worked in the same city and the first WhatsApp group of workers, facilitating our encounters with its members as researchers, but also as his friends. Through this group, we discovered the existence of other similar groups, both nationally and locally, in other cities. We started asking to join the workers' private groups in July 2020. We transparently explained our research agenda and introduced ourselves as researchers in each group. Once we had been admitted by the administrators, we clarified that some of them might be invited to an interview at a later stage. Admittance was granted to us thanks to prior acquaintance with some of the chat group members, who worked as trusted intermediaries.

Within a couple of months, we were part of dozens of WhatsApp groups of Italian couriers, through which we recruited the first interviewees, who in turn introduced us to other courier friends of theirs. We then extended this research method to other countries as well. The researchers who worked with us on the AlgoRes project were all trained through an in-house workshop and replicated this model in their respective fieldwork. The interviewees agreed to participate under the condition of full anonymization and protection of the interview data. Screenshots of images posted by couriers in WhatsApp and WeChat groups were taken only after obtaining consent from the authors of the messages. Further, we discussed with our interviewees which algorithmic tactics could be shared and which were better kept hidden because they could directly or indirectly pose a threat to the workers. Data originated from the project were shared by members of the team in a safe, encrypted online environment. It was made clear from the start that each researcher could use the data gathered through their research, so long as the co-PIs were informed. This allowed researchers to foster their careers by using their data in their theses, talks, and academic and nonacademic publications.

The results of this project were presented at several conferences and workshops, which allowed us to refine the concepts and improve our understanding of the data.[9] Part of this work has been published in articles focusing on Chinese riders' resistance strategies[10] and on how instant messaging apps afford learning, resistance, and solidarity among food delivery workers.[11]

GAMING CULTURE

The data on Instagram engagement groups are the result of a digital ethnography that Tiziano Bonini and his research assistant, Francesca Murtula, carried out from October 2019 to May 2020. Daily nonparticipant observation was complemented with twelve semistructured interviews to Instagram microinfluencers between March 11 and 25, 2020. Digital ethnography consisted in the observation of several online groups (on Facebook, Telegram, and WhatsApp), the continuous generation of a series of field notes, the transcription and open-coding of semistructured interviews with group members, and the collection of screenshots of online chats. The first engagement groups were intercepted by searching for keywords such as "pod," "engagement group," and "like exchange" on Facebook and Telegram, while other groups were added to the field after our interviewees reported them. Semistructured interviews were recorded with the consent of the respondents and lasted from a minimum of thirty minutes to a maximum of one hour; after the first interview, all respondents were contacted several times for further investigation, and we conducted a second interview with four of them.

The twelve interviewees (six women and six men, with an average age between eighteen and thirty years, with experience in pods ranging from a few weeks to a few years) defined themselves as Instagram microinfluencers; they derive an advantage (albeit minimal) from their work on the platform, but none of them live off the income from this work alone. The number of their followers ranges from 2,000 to a maximum of 25,600. Part of this work has been published in an Italian scientific journal.[12] Francesca Murtula's research was financed through a research fellowship granted by the Department of Social, Political, and Cognitive Sciences at the University of Siena.

GAMING POLITICS

The research on gaming politics, partly builds on previous projects that Emiliano either coordinated or was involved in. Reflections on Mexican and Latin American social movements and algorithmic activism are based on three research projects where Emiliano acted as PI. These projects have received funding from the 2012 Mexican Faculty Improvement Program (fund number 103.5/12/3667 and professor number UAQ-PTC-224), the FOFI-UAQ Fund 2012 (project number FCP201206) and the FOFI-UAQ-Fund 2013 (project number FCP201410) of the Autonomous University of Querétaro. Insights on the 15M movement in Spain were gathered in the context of a project supported by an Insight Development Grant from the Social Sciences and Humanities Research Council of Canada (file number 430-2014-00181). Data from these projects were assembled in a new database on NVivo and reanalyzed in a constant dialogue with the research aims of this book and a focus on the dynamics of strategic and tactical algorithmic politics and their interplay. This included the exploration of a corpus of fifty-six interviews, eighty-three documents produced by institutions and activists (leaflets, posters, social media posts and videos on various platforms, digital images, and others) complemented by extended ethnographic notes during short periods of fieldwork from 2012 to 2018. Funding from the School of Journalism, Media, and Culture, Cardiff University, supported the work of Thomas Davis, a research assistant who gathered and analyzed a corpus of press and academic articles around algorithmic politics that were added to the general database. Thomas also prepared a preliminary report with key insights into the unfolding of algorithmic practices in politics in different countries around the world, for which we are extremely grateful. To add to these data, seven more interviews were carried out by Emiliano with prominent algorithmic activists in Mexico, Italy, and Spain during 2020. Further, additional data were gathered on the algorithmic tactics of the Italian far-right through observation, interaction, and message exchanges on Twitter. Some of the reflections of this research have been published in a social movements studies' journal.[13]

DIGITAL, ALGORITHMIC, OR HYBRID ETHNOGRAPHY?

We are hardly the first social scientists to adapt our methods to the pandemic. Dozens of articles have come out on how to do research during lockdowns, and many have focused on how to explore ethnography without being able to move from home or from their city.[14] In any case, there was no need for a pandemic to digitize ethnography. Digital ethnography was an established research method well before the outburst of COVID-19.[15] As the sociologist Angèle Christin points out, "It is important to acknowledge that such a turn is also far from new."[16] However, Christin adds that qualitative scholars analyzing digital data should be more aware of the question of "how do software and platforms shape exchanges and representations,"[17] and they should engage in what she calls "algorithmic ethnography,"[18] a way of doing digital ethnography with a specific focus on the "study of the computational systems enabling and shaping online interactions."[19] Without even being aware of it, our way of working with digital data for this book was partly aligned with the concept of "algorithmic ethnography" developed by Christin.

The research that fuels this book, then, is the result of several digital ethnography research projects, complemented by more traditional ethnographic practices of face-to-face interviews and periods of participant observation, both online and offline. Both digital and traditional ethnographies are multisited.[20] We define our fieldwork as multisited and hybrid not only because we have generated data in different countries, through the collaboration of an extensive network of researchers, but also because we have practiced long periods of participant observation on different digital fields, such as dozens of private WhatsApp, Telegram, and Facebook groups. The ethnographic field in this case weaves together intense networks of human and nonhuman actors, platforms, conversations, audio messages, videos, emails, face-to-face meetings, and platforms. As the anthropologist Gabriele de Seta argues, the meaning of "being there" (in the field) evoked by Ulf Hannerz[21] is clearly different in this context: "Being there on different platforms and services, different conversations and groups, updated and in-the-loop regarding different topics and events: the spatial experience of the Internet was way more social than technological."[22] What then has become the famous "ethnographic field"? As de Seta rightly notes, the field is a network, and the ethnographer is a "networked field weaver."[23]

Scholars have pointed out that online research has mostly been carried out in English since "most research on the Internet is centred in Anglo-American cultural contexts."[24] However, according to Clare Madge, this can restrict the agency of people who do not speak this idiom as a first language to express their views.[25] By mainly relying on local, multilingual researchers, our aim was to mitigate this aspect promoting digital sociological inquiry beyond the Western world.

WHATSAPP AND THE POWER OF SCREENSHOTS

In this extended field, which blurs the boundaries between online and offline worlds, ethnographic materials are no longer just the physical objects that ethnographers would normally take home as indices of the culture they were studying—they can also be digital objects such as memes, emojis, videos, and audio messages posted by users in a Facebook or WhatsApp group. The anthropologists Edgar Gómez-Cruz and Ignacio Siles documented the relevance of visual elements such as emojis, stickers, and GIFs in the daily use of WhatsApp in Mexico City.[26] These visual elements also proved to be central to our observations. Through them, group users conveyed moods and opinions. Some of these images then became part of the specific jargon that developed within a chat, acquiring new meanings. We can therefore consider them to be, to all intents and purposes, typical expressions of the culture of a social group interacting through WhatsApp. Thus, our ethnographic materials include not only the photos of food delivery riders that we took when we met them face-to-face, but also the hundreds of screenshots of textual conversations and visual elements that we collected in the online groups we frequented for more than a year. As Jan Švelch noted, screenshots are "cultural artifacts."[27] These screenshots were then discussed in our daily chat dialogues. It often happened that one of us would share a screenshot in our private chat so we could interpret it together.[28] As Deborah Lupton pointed out, the benefits of considering these groups as fieldwork are the "thick data and the prolonged engagement of participants."[29]

Of all the groups that we observed on the different platforms, those on WhatsApp were the ones we frequented the most, which allowed us to better understand the worlds that we were studying. WhatsApp is a new object of study for media scholars but, as Gómez-Cruz and Harindranath argue, "WhatsApp is [a] quintessential research object if we want to understand digital culture(s) from a non-data-logic standpoint, using a non-media-centric perspective, especially in the Global South."[30] WhatsApp and similar devices have become part of people's communication routines and are tools through which people express their culture, and as cultural studies researchers, we cannot help but include them in our fieldwork. This strategy has allowed us to resolve some of the limitations caused by social distancing measures and our initial inability to access the fields and populations we wanted to study.

WhatsApp's role in the research process for this book has been twofold: (1) as a field notebook and (2) as fieldwork. It was both the field and the place to reflect on the fieldwork. In other words, WhatsApp has constituted both the medium (the field) and the metamedium (the field notebook) of our research. However, our toolbox did not stop at digital ethnography. In the intermittent periods between lockdowns, we also managed to practice some more traditional ethnography, meeting our informants in the field, observing them during their daily activities, and interviewing them face-to-face, as well as on Zoom, GMeet, Skype, and Jitsi.

DESK RESEARCH AND OTHER DATA

Alongside our ethnographic work, we carried out desk research on both scientific papers and the press around algorithm gaming practices. Between 2020 and 2021, we built a database with 100 press articles and 150 scientific articles. We did a content analysis of this database and coded the press articles according to three categories: gaming culture, politics, and gig work, while the scientific articles were coded both according to their relevance to one of these three categories and their adherence to the themes of agency and resistance. All the data generated through the various ethnographies and the content analysis are the result of an iterative process inspired by grounded theory.[31] We always entered the field with general questions, driven by curiosity to find out "What's going on here?"

Some of the data mentioned marginally in this book come from other research projects that we coordinated or were involved in. The data on the gaming tactics of Airbnb hosts come from the article "'Ancora non ci ho capito niente di come funziona l'algoritmo': La consapevolezza algoritmica degli host di Airbnb," ("'I Still Don't Understand How the Algorithm Works': The Algorithmic Awareness of Airbnb Hosts"), coauthored by Francesca Murtula and Tiziano Bonini.[32] Finally, the tactics from Tinder come from the master's thesis in communication and media studies of Susanna Bonelli, supervised by Tiziano Bonini at the University of Siena.

NOTES

INTRODUCTION

1. Stefano is the fictional name of our informant. We changed his real name to protect his identity.

2. Krishnan Vasudevan and Ngai Keung Chan, "Gamification and Work Games: Examining Consent and Resistance among Uber Drivers," *New Media & Society* 24, no. 4 (2022): 866–886.

3. Ulises A. Mejias and Nick Couldry, "Datafication," *Internet Policy Review* 8, no. 4 (2019): 1–10.

4. Colin Koopman, *How We Became Our Data* (Chicago: University of Chicago Press, 2019).

5. Nick Couldry and Ulises A. Mejias, *The Costs of Connection* (Stanford, CA: Stanford University Press, 2019).

6. Shoshana Zuboff, *The Age of Surveillance Capitalism* (New York: Public Affairs, 2019).

7. José van Dijck, David B. Nieborg, and Thomas Poell, "Reframing Platform Power," *Internet Policy Review* 8, no. 2 (2019): 1–18.

8. Virginia Eubanks, *Automating Inequality: How High-Tech Tools Profile, Police, and Punish the Poor* (New York: St. Martin's Press, 2018).

9. From 2013 to 2019, the Dutch government used a secret algorithm to detect possible tax fraud. This system, called SyRI (for "system risk indication"), calculated the risk profiles of residents who were supposedly more likely to commit fraud, analyzed the risk profiles of residents who claimed refunds for child expenses (such as daycare fees), and flagged likely fraudsters, who were then forced to pay money that they did not actually owe. As a result, the system ordered 26,000 Dutch families (mostly poor

and mixed- or dual-nationality families) to repay thousands of euros in benefits that the system said they received unfairly. See Gabriel Geiger, "How a Discriminatory Algorithm Wrongly Accused Thousands of Families of Fraud," Vice.com, March 2, 2021, https://medium.com/vice/how-a-discriminatory-algorithm-wrongly-accused-th ousands-of-families-of-fraud-94026788d41f; Jordan Pearson, "The Story of How the Australian Government Screwed Its Most Vulnerable People," Motherboard, August 24, 2020, https://www.vice.com/en/article/y3zkgb/the-story-of-how-the-australian-govern ment-screwed-its-most-vulnerable-people-v27n3.

10. Alessandro Gandini, "Digital Labour: An Empty Signifier?" *Media, Culture & Society* 43, no. 2 (2021): 371–372.

11. Gandini, "Digital Labour: An Empty Signifier?" 375.

12. Tiziana Terranova, "Free Labor: Producing Culture for the Digital Economy" *Social Text* 18, no. 2 (2000): 33–58; Christian Fuchs, "Labor in Informational Capitalism and on the Internet," *The Information Society* 26, no. 3 (2010): 179–196; Trebor Scholz, ed., *Digital Labor: The Internet as Playground and Factory* (New York: Routledge, 2012).

13. Dallas Smythe, "On the Audience Commodity and Its Work," in *Dependency Road: Communications, Capitalism, Consciousness, and Canada*, ed. Dallas Smythe (Norwood, NJ: Ablex, 1981), 22–51.

14. Gandini, "Digital Labour: An Empty Signifier?" 375.

15. Gandini, "Digital Labour: An Empty Signifier?" 375.

16. Alex Rosenblat and Luke Stark, "Algorithmic Labor and Information Asymmetries: A Case Study of Uber's Drivers," *International Journal of Communication* 10, no. 27 (2016): 3758–3784; Niels van Doorn, "Platform Labor: On the Gendered and Racialized Exploitation of Low-Income Service Work in the 'On-Demand' Economy," *Information, Communication & Society* 20, no. 6 (2017): 898–914.

17. Ien Ang, "Culture and Communication: Towards an Ethnographic Critique of Media Consumption in the Transnational Media System," *European Journal of Communication* 5, no. 2 (1990): 239–260.

18. Ang, "Culture and Communication," 247.

19. Anthony Giddens, *The Constitution of Society. Outline of the Theory of Structuration* (London: Polity, 1984).

20. Jocelyn Hollander and Rachel Einwohner, "Conceptualizing Resistance," *Sociological Forum* 19, no. 4 (2004): 533–554.

21. James C. Scott, *Weapons of the Weak. Everyday Forms of Peasant Resistance* (New Haven, CT: Yale University Press, 1985).

22. Michel de Certeau, *The Practice of Everyday Life* (Berkeley: University of California Press, 1984).

23. Edward P. Thompson, "The Moral Economy of the English Crowd in the Eighteenth Century," *Past & Present* 22, no. 50 (1971): 76–136; James C. Scott, *The Moral*

Economy of the Peasant: Rebellion and Subsistence in Southeast Asia (New Haven, CT: Yale University Press, 1976).

24. Mark Fisher, *Capitalist Realism: Is There No Alternative?* (London: John Hunt Publishing, 2009).

25. Francesca Antonini, "Pessimism of the Intellect, Optimism of the Will: Gramsci's Political Thought in the Last Miscellaneous Notebooks," *Rethinking Marxism* 31, no. 1 (2019): 42–57.

26. Caitlin Petre, Brooke E. Duffy, and Emily Hund, "'Gaming the System': Platform Paternalism and the Politics of Algorithmic Visibility," *Social Media+Society* 5, no. 4 (October 2019). https://doi.org/10.1177/2056305119879995.

27. Howard S. Becker, "Whose Side Are We On?" *Social Problems* 14, no. 3 (1967): 239–247.

CHAPTER 1

1. Lisa Parks and Nicole Starosielski, eds., *Signal Traffic: Critical Studies of Media Infrastructures* (Champaign: University of Illinois Press, 2015); John Durham Peters, *The Marvelous Clouds: Toward a Philosophy of Elemental Media* (Chicago: University of Chicago Press, 2015); Jean Christophe Plantin, Carl Lagoze, Paul Edwards, and Christian Sandvig, "Infrastructure Studies Meet Platform Studies in the Age of Google and Facebook," *New Media & Society* 20, no. 1 (2018): 293–310.

2. Martin Moore and Damian Tambini, eds., *Digital Dominance: The Power of Google, Amazon, Facebook, and Apple* (Oxford: Oxford University Press, 2018).

3. Marc Steinberg, Rahul Mukherjee, and Aswin Punathambekar, "Media Power in Digital Asia: Super Apps and Megacorps," *Media, Culture & Society* (2022): 01634437221127805; Michael Keane and Anthony Fung, "Digital Platforms: Exerting China's New Cultural Power in the Asia-Pacific," *Media Industries* 5, no. 1 (2018): 47–50.

4. José van Dijck, Thomas Poell, and Martijn de Waal, *The Platform Society: Public Values in a Connective World* (Oxford: Oxford University Press, 2018). David B. Nieborg and Thomas Poell, "The Platformization of Cultural Production: Theorizing the Contingent Cultural Commodity," *New Media & Society* 20, no. 11 (2018): 4275–4292; Nick Srnicek, *Platform Capitalism* (London: Polity, 2017).

5. Plantin et al., "Infrastructure Studies Meet Platform Studies in the Age of Google and Facebook,"; Jean Christophe Plantin and Aswin Punathambekar, "Digital Media Infrastructures: Pipes, Platforms, and Politics," *Media, Culture & Society* 41, no. 2 (2019): 163–174.

6. Moore and Tambini, *Digital Dominance*.

7. Ganaele Langlois and Greg Elmer, "The Research Politics of Social Media Platforms," *Culture Machine* 14 (2013), http://www.culturemachine.net/index.php/cm/article/view/505.

8. Zuboff, *The Age of Surveillance Capitalism*.

9. David Hesmondhalgh, "The Infrastructural Turn in Media and Internet Research," in *Routledge Companion to Media Industries*, ed. Paul McDonald (London: Taylor & Francis, 2021), 132–142.

10. Tarleton Gillespie, Pablo Boczkowski, and Kirsten Foot, *Media Technologies: Essays on Communication, Materiality, and Society* (Cambridge, MA: MIT Press, 2014), 221.

11. John Hondros, "The Internet and the Material Turn,." *Westminster Papers in Communication and Culture* 10, no. 1 (2015): 1–3; Nathalie Casemajor, "Digital Materialisms: Frameworks for Digital Media Studies," *Westminster Papers in Communication and Culture* 10, no. 1 (2015): 4–17.

12. Durham Peters, *The Marvelous Clouds*.

13. Harold A. Innis, *The Bias of Communication* (Toronto: University of Toronto Press, 1982); Walter Ong, *Orality and Literacy: The Technologizing of the Word* (London and New York: Taylor and Francis, 1982). Marshall McLuhan, *Understanding Media: The Extensions of Man* (Cambridge, MA: MIT Press, 1964).

14. Cathy O'Neil, *Weapons of Math Destruction: How Big Data Increases Inequality and Threatens Democracy* (New York: Crown, 2016); Eubanks, *Automating Inequality*; Sofya U. Noble, *Algorithms of Oppression: How Search Engines Reinforce Racism* (New York: New York University Press, 2018); Ruha Benjamin, *Race after Technology* (London: Polity, 2019); Abeba Birhane, "Algorithmic Injustice: A Relational Ethics Approach," *Patterns* 2, no. 2 (2021): 100205; Sasha Costanza-Chock, *Design Justice: Community-led Practices to Build the Worlds We Need* (Cambridge, MA: MIT Press, 2020).

15. Nick Couldry and Ulises Mejias, "The Decolonial Turn in Data and Technology Research: What Is at Stake and Where Is It Heading?" *Information, Communication & Society* (2021): 786. See also Lina Dencik, Arne Hintz, Joanna Redden, and Emiliano Treré, *Data Justice* (London: SAGE, 2022).

16. Nick Couldry and Ulises Mejias, *The Costs of Connection* (Stanford, CA: Stanford University Press, 2019).

17. Dencik et al., *Data Justice*.

18. Jathan Sadowski, *Too Smart: How Digital Capitalism Is Extracting Data, Controlling Our Lives and Taking Over the World* (Cambridge, MA: MIT Press, 2020).

19. Zuboff, *The Age of Surveillance Capitalism*.

20. Zuboff, *The Age of Surveillance Capitalism*, 378.

21. Sonia Livingstone, "Audiences in an Age of Datafication: Critical Questions for Media Research," *Television & New Media* 20, no. 2 (2019): 170–183.

22. Marit de Jong and Robert Prey, "The Behavioral Code: Recommender Systems and the Technical Code of Behaviorism," in *The Necessity of Critique*, ed. Darryl Cressman (Cham, Switzerland: Springer Cham 2022), 143–159.

23. See de Jong and Prey, "The Behavioral Code," 143.

24. See de Jong and Prey, "The Behavioral Code," 143.

25. Nancy Ettlinger, "Algorithmic Affordances for Productive Resistance," *Big Data & Society* 5, no. 1 (2018): 4.

26. Nick Couldry and Alison Powell, "Big Data from the Bottom Up," *Big Data & Society* 1, no. 2 (2014): 2053951714539277.

27. Stefania Milan, "Data Activism as the New Frontier of Media Activism," in *Media Activism in the Digital Age*, eds. Goubin Yang and Viktor Pickard (London: Routledge, 2017): 151–163.

28. Helen Kennedy, "Living with Data: Aligning Data Studies and Data Activism through a Focus on Everyday Experiences of Datafication," *Krisis: Journal for Contemporary Philosophy* 1 (2018): 18–30. See also Jean Burgess, Kate Albury, Anthony McCosker, and Rowan Wilken, *Everyday Data Cultures* (London: Polity, 2022); Ignacio Siles, *Living with Algorithms: Agency and User Culture in Costa Rica* (Cambridge, MA: MIT Press, 2023).

29. See van Dijck et al., "Reframing Platform Power."

30. Michel Foucault, *The History of Sexuality. Vol. I* (London: Penguin Books, 1978), 95–96.

31. Taina Bucher, "The Algorithmic Imaginary: Exploring the Ordinary Affects of Facebook Algorithms," *Information, Communication & Society* 20, no. 1 (2017): 42.

32. Julia Velkova and Anne Kaun, "Algorithmic Resistance: Media Practices and the Politics of Repair," *Information, Communication & Society* 24, no. 4 (2021): 523–540.

33. Jeremy W. Morris, "Gaming Music Platforms," paper presented at the Platformization of Cultural Production Workshop, October 8, 2018, University of Toronto.

34. Rob Kitchin, "Thinking Critically about and Researching Algorithms," *Information, Communication & Society* 20, no. 1 (2017): 26.

35. Giddens, *The Constitution of Society* (London: Polity, 1984), 14.

36. Nick Couldry, "Inaugural: A Necessary Disenchantment: Myth, Agency and Injustice in a Digital World," *The Sociological Review* 62, no. 4 (2014): 880–897.

37. Till Jansen, "Who Is Talking? Some Remarks on Nonhuman Agency in Communication," *Communication Theory* 26 (2016): 255–272.

38. Massimo Airoldi, *Machine Habitus: Toward a Sociology of Algorithms* (London: Polity, 2021).

39. Gina Neff and Peter Nagy, "Automation, Algorithms, and Politics Talking to Bots: Symbiotic Agency and the Case of Tay," *International Journal of Communication* 10, no. 17 (2016): 4915–4931.

40. Jansen, "Who Is Talking? Some Remarks on Nonhuman Agency in Communication."

41. Giddens, *The Constitution of Society*.

42. Giddens, *The Constitution of Society*, 374.

43. Foucault, *The History of Sexuality*, 95.

44. George Simmel, *The Sociology of Georg Simmel* (Glencoe, IL: The Free Press, 1950), quoted in David Courpasson and Steven Vallas, "Resistance Studies: A Critical Introduction," in *Sage Handbook of Resistance* (London: SAGE, 2016), 1–28.

45. Ian Buchanan, *Michel de Certeau: Cultural Theorist* (London: SAGE, 2000), 89.

46. Courpasson and Vallas, "Resistance Studies: A Critical Introduction," 5.

47. Brent L. Pickett, "Foucault and the Politics of Resistance," *Polity* 28, no. 4 (1996): 445–466.

48. Hollander and Einwohner, "Conceptualizing Resistance."

49. We are aware of the controversy about the collaboration of James Scott with the US Central Intelligence Agency (CIA) in Indonesia during the 1960s. A tweet posted on Twitter on February 13, 2022, reignited this news: https://twitter.com /alp1111112/status/1492869548596506626?s=20&t=3eykrOPO4s7notAD3rSjwQ. Scott was not the only American social scientist to collaborate with the CIA, as this detailed study by David Price demonstrates. See David Price, *Cold War Anthropology: The CIA, the Pentagon, and the Growth of Dual Use Anthropology* (Durham, NC: Duke University Press, 2016). The relationship between Scott and the CIA is part of a much larger debate involving Western social sciences and their ambiguous relationships with power. In spite of this, we still consider Scott's scientific contributions significant, even if they will rightly have to be reread in the light of these recent discoveries about his past.

50. Scott, *Weapons of the Weak*.

51. Courpasson and Vallas, "Resistance Studies: A Critical Introduction."

52. Courpasson and Vallas, "Resistance Studies: A Critical Introduction," 5.

53. Scott, *Weapons of the Weak*, 6.

54. Mikael Baaz, Mona Lilja, Michael Schulz, and Stellan Vinthagen, "Defining and Analyzing 'Resistance': Possible Entrances to the Study of Subversive Practices," *Alternatives* 41, no. 3 (2016): 137–153.

55. Baaz et al., "Defining and Analyzing 'Resistance,'" 142.

56. Emiliano Treré, "From Digital Activism to Algorithmic Resistance," in *Routledge Companion to Media and Activism*, ed. Graham Meikle (London and New York: Routledge, 2018), 367–375.

57. Davide Beraldo and Stefania Milan, "From Data Politics to the Contentious Politics of Data," *Big Data & Society* 6, no. 2 (2019): 1–11.

58. Beraldo and Milan, "From Data Politics to the Contentious Politics of Data," 6.

59. Beraldo and Milan, "From Data Politics to the Contentious Politics of Data," 6.

60. Mimi Onuoha and Mother Cyborg, *A People's Guide to AI*, (Allied Media, 2018), https://alliedmedia.org/wp-content/uploads/2020/09/peoples-guide-ai.pdf.

61. Ettlinger, "Algorithmic Affordances for Productive Resistance," 4.

62. Ettlinger, "Algorithmic Affordances for Productive Resistance," 5.

63. Ettlinger, "Algorithmic Affordances for Productive Resistance," 5.

64. Scott, *Weapons of the Weak*, 29.

65. Scott, *Weapons of the Weak*, 29.

66. Roger Silverstone, "The Power of the Ordinary: On Cultural Studies and the Sociology of Culture," *Sociology* 28, no. 4 (1994): 991–1001.

67. See de Certeau, *The Practice of Everyday Life*.

CHAPTER 2

1. The concept of the moral economy was first popularized by the work of E. P. Thompson (1971), and then by the work of James Scott (1976). See Thompson, "The Moral Economy of the English Crowd in the Eighteenth Century"; Scott, *The Moral Economy of the Peasant: Rebellion and Subsistence in Southeast Asia*.

2. Alessandro Manzoni, *The Betrothed* (London: Richard Bentley, 1834), 141.

3. Manzoni, *The Betrothed*, 153.

4. Edward P. Thompson, *Customs in Common: Studies in Traditional Popular Culture* (New York: New Press, 1993), 185.

5. Jaime Palomera and Theodora Vetta, "Moral Economy: Rethinking a Radical Concept," *Anthropological Theory* 16, no. 4 (2016): 413–432.

6. Thompson, "The Moral Economy of the English Crowd in the Eighteenth Century," 79.

7. Thompson, *Customs in Common*, 188.

8. See also David Harvie and Keir Milburn, "The Moral Economy of the English Crowd in the Twenty-First Century," *South Atlantic Quarterly* 112, no. 3 (2013): 559–567.

9. Andrew Charlesworth and Adrian Randall, "Morals, Markets and the English Crowd in 1766," *Past & Present* 114 (1987): 200–213.

10. Thompson, *Customs in Common*, 273.

11. Andrew Sayer, "Valuing Culture and Economy," in *Culture and Economy after the Cultural Turn*, eds. Larry Ray and Andrew Sayer (London: SAGE, 1999), 68.

12. Marc Edelman, "E. P. Thompson and Moral Economies," in *A Companion to Moral Anthropology*, ed. Didier Fassin (London: Whiley-Blackwell, 2012): 49–66.

13. Thompson, "The Moral Economy of the English Crowd in the Eighteenth Century," 135.

14. Langdon Winner, "Do Artifacts Have Politics?" *Daedalus* 109, no. 1 (1980): 121–136.

15. Anne Jorunn Berg and Merete Lie, "Feminism and Constructivism: Do Artifacts Have Gender?," *Science, Technology, & Human Values* 20, no. 3 (1995): 332–351.

16. Tarleton Gillespie, *Custodians of the Internet* (New Haven, CT: Yale University Press, 2018).

17. Louise Amoore, *Cloud Ethics: Algorithms and the Attributes of Ourselves and Others* (Durham, NC: Duke University Press, 2020), 7.

18. Jenna Burrell and Marion Fourcade, "The Society of Algorithms," *Annual Review of Sociology* 47 (2021): 227.

19. See Hebrew University of Jerusalem, "Digital Values: The Construction of Values in Digital Spheres," https://digitalvalues.huji.ac.il/.

20. Blake Hallinan, Rebecca Scharlach, and Limor Shifman, "Beyond Neutrality: Conceptualizing Platform Values," *Communication Theory* 32, no. 2 (2022): 201–222.

21. Hallinan, Scharlach, and Shifman, "Beyond Neutrality," 204.

22. Bruno Latour, "Technology Is Society Made Durable," *Sociological Review* 38, no. 1_Suppl (1990): 103–131.

23. Tarleton Gillespie, "Do Not Recommend? Reduction as a Form of Content Moderation," *Social Media + Society* 8, no. 3 (2022): 20563051221117552.

24. See van Dijck et al., *Platform Society*.

25. Tiziano Bonini and Eleonora M. Mazzoli, "A Convivial-agonistic Framework to Theorise Public Service Media Platforms and Their Governing Systems," *New Media & Society* 24, no. 4 (2022): 922–941.

26. Hallinan, Scharlach, and Shifman, "Beyond Neutrality," 217.

27. Jodi Dean, *Democracy and Other Neoliberal Fantasies* (Durham, NC: Duke University Press, 2009).

28. Gillespie, *Custodians of the Internet*.

29. Jonathan Cohn, *The Burden of Choice: Recommendations, Subversion, and Algorithmic Culture* (New Brunswick, NJ: Rutgers University Press, 2019).

30. Peter Guest and Youyou Zhou, "The Gig Workers Index: Mixed Emotions, Dim Prospects," *Rest of World*, September 21, 2021, https://restofworld.org/2021/global-gig-workers-index-mixed-emotions-dim-prospects/.

31. Cheryll Ruth Soriano and Jason Vincent Cabañes, "Entrepreneurial Solidarities: Social Media Collectives and Filipino Digital Platform Workers," *Social Media + Society* 6, no. 2 (2020): 2056305120926484.

32. Petre, Duffy, and Hund, "'Gaming the System.'"

33. Tarleton Gillespie, "Algorithmically Recognizable: Santorum's Google Problem, and Google's Santorum Problem," *Information, Communication & Society* 20, no. 1 (2017): 63–80; Malte Ziewitz, "Rethinking Gaming: The Ethical Work of Optimization in Web Search Engines," *Social Studies of Science* 49, no. 5 (2019): 707–731.

34. Thomas Poell, David Nieborg, and Brooke E. Duffy, *Platforms and Cultural Production* (London: Polity, 2022), 103.

35. Poell, Nieborg, and Duffy, *Platforms and Cultural Production*, 103.

36. Petre, Duffy, and Hund, "'Gaming the System,'" 2.

37. Petre, Duffy, and Hund, "'Gaming the System,'" 2.

38. Petre, Duffy, and Hund, "'Gaming the System,'" 2.

39. Jereny W. Morris, "Music Platforms and the Optimization of Culture," *Social Media + Society* 6, no. 3 (July 2020). https://doi.org/10.1177/2056305120940690.

40. Ziewitz, "Rethinking Gaming," 723.

41. David Hesmondhalgh, "Cultural Studies, Production and Moral Economy," *Reseaux* 4, no. 192 (2015): 169–202. See also Joshua Green and Henry Jenkins, "The Moral Economy of Web 2.0: Audience Research and Convergence Culture," in *Media Industries: History, Theory and Method*, eds. Jennifer Holt and Alisa Perren (Malden, MA: Wiley-Blackwell, 2009), 213–225.

42. David Hesmondhalgh, "Capitalism and the Media: Moral Economy, Well-being and Capabilities," *Media, Culture & Society* 39, no. 2 (2017): 202–218.

43. Henry Jenkins, Sam Ford, and Joshua Green, *Spreadable Media: Creating Value and Meaning in a Networked Culture* (New York: New York University Press, 2013).

44. Roger Silverstone, Eric Hirsch, and David Morley, "Information and Communication Technologies and the Moral Economy of the Household," in *Consuming Technologies: Media and Information in Domestic Spaces*, eds. Roger Silverstone and Eric Hirsch (London: Routledge, 1992), 22.

45. On the value of visibility in the platformized cultural industries, see chapter 5 of Poell, Nieborg, and Duffy, *Platforms and Cultural Production*.

46. Stine Lomborg and Patrick H. Kapsch, "Decoding Algorithms," *Media, Culture & Society* 42, no. 5 (2020): 745–761.

47. Stuart Hall, "Encoding and Decoding in the Television Discourse," in *CCCS Selected Working Papers*, eds. Ann Gray et al. (London: Routledge, 2007), 402–414.

48. Ignacio Siles, Amy Ross Arguedas, Mónica Sancho, and Ricardo Solís-Quesada, "Playing Spotify's Game: Artists' Approaches to Playlisting in Latin America," *Journal of Cultural Economy* (2022). https://doi.org/10.1080/17530350.2022.2058061.

49. Mohammad A. Anwar and Mark Graham, "Hidden Transcripts of the Gig Economy: Labour Agency and the New Art of Resistance among African Gig Workers," *Environment and Planning A: Economy and Space* 52, no. 7 (October 2020): 1269–1291. https://doi.org/10.1177/0308518X19894584.

50. Ziewitz, "Rethinking Gaming," 723.

51. See de Certeau, *The Practice of Everyday Life*.

52. See de Certeau, *The Practice of Everyday Life*, 35–36.

53. See de Certeau, *The Practice of Everyday Life*, 36–37.

54. Buchanan, *Michel de Certeau*, 89.

55. Velkova and Kaun, "Algorithmic Resistance: Media Practices and the Politics of Repair."

56. Justine Gangneux, "Tactical Agency? Young People's (Dis)engagement with WhatsApp and Facebook Messenger," *Convergence* 27, no. 2 (April 2021): 458–471. https://doi.org/10.1177/1354856520918987.

57. Tanya Kant, *Making It Personal: Algorithmic Personalization, Identity, and Everyday Life* (Oxford: Oxford University Press, 2020), 215.

58. Samuel C. Woolley and Phil Howard, eds., *Computational Propaganda: Political Parties, Politicians, and Political Manipulation on Social Media* (Oxford: Oxford University Press, 2018).

59. Taina Bucher, "The Algorithmic Imaginary: Exploring the Ordinary Affects of Facebook Algorithms," *Information, Communication & Society* 20, no.1 (2017): 30–44.

60. Bucher, "The Algorithmic Imaginary," 42.

61. Jeremy W. Morris, Robert Prey, and David B. Nieborg, "Engineering Culture: Logics of Optimization in Music, Games, and Apps," *Review of Communication* 21, no. 2 (2021):161–175.

62. Lauren Andres, Phil Jones, Stuart Paul Denoon-Stevens, and Melgaço Lorena, "Negotiating Polyvocal Strategies: Re-reading de Certeau through the Lens of Urban Planning in South Africa," *Urban Studies* 57, no. 12 (September 2020): 2440–2455. https://doi.org/10.1177/0042098019875423.

63. Andres, Jones, Denoon-Stevens, and Lorena, "Negotiating Polyvocal Strategies," 2444.

64. Andres, Jones, Denoon-Stevens, and Lorena, "Negotiating Polyvocal Strategies," 2444.

65. The vignettes that provide a concrete example of these four manifestations of agency come from the stories that we collected through desk analyses, interviews, participant observations in digital environments, and informal conversations with key informants. The characters described are fictitious (apart from the Mexican example), but they represent examples of real practices.

66. Nieborg and Poell, "The Platformization of Cultural Production."

67. Morris, Prey, and Nieborg, "Engineering Culture: Logics of Optimization in Music, Games, and Apps," 162.

68. Morris, Prey, and Nieborg, "Engineering Culture: Logics of Optimization in Music, Games, and Apps," 162.

69. Tiziano Bonini and Alessandro Gandini, "'First Week Is Editorial, Second Week Is Algorithmic': Platform Gatekeepers and the Platformization of Music Curation," *Social Media + Society* (October 2019), 9. https://doi.org/10.1177/2056305119880006.

70. Jeremy W. Morris, "Music Platforms and the Optimization of Culture," *Social Media + Society* 6, no. 3 (July 2020), 7, https://doi.org/10.1177/2056305120940690.

71. Tiziano Bonini and Francesca Murtula, "'Ancora non ci ho capito niente di come funziona l'algoritmo': La consapevolezza algoritmica degli host di Airbnb" ("'I Still Don't Understand How the Algorithm Works': Airbnb Hosts' Algorithmic Awareness"), *Sociologia Italiana* 19–20 (2022): 147–161.

72. Bonini and Murtula, "'Ancora non ci ho capito niente di come funziona l'algoritmo,'" 156.

73. This is what Taina Bucher calls "algorithmic imaginaries" and can be shaped both by personal experience and by conversations with other users. These conversations are called "algorithmic gossip" by Sophie Bishop. See Bucher, "The Algorithmic

Imaginary," and Sophie Bishop, "Managing Visibility on YouTube through Algorithmic Gossip," *New Media & Society* 21, nos. 11–12 (2019): 2589–2606.

74. Pandemia Digital, "Alfredo del Mazo, Gobernador del Estado de México, Usa Más de 60.000 Bots," Pandemia Digital.net, June 29, 2020, https://www.pandemiadigital.net/bots/alfredo-del-mazo-gobernador-del-estado-de-mexico-usa-mas-de-60-000-bots/.

75. Marina Ayeb and Tiziano Bonini "'It Was Very Hard for Me to Keep Doing That Job': Understanding Troll Farm's Working in the Arab world," *Social Media + Society*, forthcoming (2023).

76. TripAdvisor, https://www.tripadvisor.com/TripAdvisorInsights/w3688.

77. TripAdvisor, https://www.tripadvisor.com/TripAdvisorInsights/w3688.

78. TripAdvisor, https://www.tripadvisor.com/TripAdvisorInsights/w3688.

79. See de Certeau, *The Practice of Everyday Life*, 39.

80. Aaron Mamiit, "Uber Drivers Reportedly Triggering Higher Fares through Surge Club," *Digital Trends*, June 17, 2019, https://www.digitaltrends.com/cars/uber-drivers-surge-club-triggers-higher-fares; Jake Shenker, "Strike 2.0. How Gig Economy Workers Are Using Tech to Strike Back," *The Guardian*, August 31, 2019, https://www.theguardian.com/books/2019/aug/31/the-new-resistance-how-gig-economy-workers-are-fighting-back.

81. Victoria O'Meara, "Weapons of the Chic: Instagram Influencer Engagement Pods as Practices of Resistance to Instagram Platform Labor," *Social Media + Society* 5, no. 4 (October 2019). https://doi.org/10.1177/2056305119879671.

82. Robert K. Merton, "The Matthew Effect in Science: The Reward and Communication Systems of Science Are Considered," *Science* 159, no. 3810 (1968): 56–63.

83. Emiliano Treré, *Hybrid Media Activism: Ecologies, Imaginaries, Algorithms* (London: Routledge, 2019).

84. For an in-depth description of the concept of sociomateriality, see Wanda Orlikowski and Susan Scott, "Sociomateriality: Challenging the Separation of Technology, Work and Organization," *Academy of Management Annals* 2, no. 1 (2008): 433–474.

85. Davide Sparti, "Tango bonding: Il tango argentino e il consumo di intimità temporanea," *Rassegna Italiana di Sociologia* 58, no. 3 (2017): 513–544.

86. On the mutual shaping of users and algorithms see Ignacio Siles, Johan Espinoza-Rojas, Adrián Naranjo, and María Fernanda Tristán, "The Mutual Domestication of Users and Algorithmic Recommendations on Netflix," *Communication, Culture & Critique* 12, no. 4 (December 2019): 499–518. https://doi.org/10.1093/ccc/tcz025.

87. See de Certeau, *The Practice of Everyday Life*.

88. Mark Andrejevic, *Automated Media* (London: Routledge, 2019), 2.

89. Zuboff, *The Age of Surveillance Capitalism*.

90. Buchanan, *Michel de Certeau*, 89–104.

CHAPTER 3

1. *The Economic Times*, "India Placed 72nd on Global List with Average Monthly Wage of Rs 32,800: Report," *The Economic Times*, August 28, 2020, https://economictimes.indiatimes.com/news/economy/indicators/india-placed-72nd-on-global-list-with-average-monthly-wage-of-rs-32800-report/articleshow/77806437.cms?utm_source=contentofinterest&utm_medium=text&utm_campaign=cppst. https://economictimes.indiatimes.com/news/economy/indicators/india-placed-72nd-on-global-list-with-average-monthly-wage-of-rs-32800-report/articleshow/77806437.cms?from=mdr.

2. DaretoDreams, "Zoomato and Swiggy Delivery Boy Earning Double with Shadowfax," YouTube, October 24, 2020. https://www.youtube.com/watch?v=mcHShFHGw7w&ab_channel=DaretoDreams (accessed July 11, 2021).

3. Swati Singh is a student at the Indraprastha College for women, Delhi University, who took part in our research project, Algorithmic Resistance. See the appendix of this book for more details.

4. The scientific literature on the subject of food delivery workers' agency is rapidly expanding. We mention here only a few authors who are relevant, especially for studies on gig working in countries of the Global South: Ya-Wen Lei, "Delivering Solidarity: Platform Architecture and Collective Contention in China's Platform Economy," *American Sociological Review* 86, no. 2 (April 2021): 279–309. https://doi.org/10.1177/0003122420979980; Ping Sun and Julie Yujie Chen, "Platform Labor and Contingent Agency in China," *China Perspectives* 1 (2021): 19–27; Chuxuan Liu and Eli Friedman, "Resistance under the Radar: Organization of Work and Collective Action in China's Food Delivery Industry," *China Journal* 86, no. 1 (July 2021): 68–89. https://doi.org/10.1086/714292; https://doi.org/10.1177/0950017019862954; Jamie Woodcock, *The Fight against Platform Capitalism: An Inquiry into the Global Struggles of the Gig Economy* (London: University of Westminster Press, 2021).

5. Richard Heeks, "Decent Work and the Digital Gig Economy: A Developing Country Perspective on Employment Impacts and Standards in Online Outsourcing, Crowdwork, etc.," *Development Informatics Working Paper* 71 (2017): 1–82, quoted in Silvia Masiero, "Digital Platform Workers under COVID-19: A Subaltern Perspective," *Proceedings of the 54th Hawaii International Conference on System Sciences* (2021): 6350.

6. Mark Graham and Jamie Woodcock, "Towards a Fairer Platform Economy: Introducing the Fairwork Foundation," *Alternate Routes* (2018): 242–253, quoted in Masiero, "Digital Platform Workers under COVID-19: A Subaltern Perspective," 6350. See also Jamie Woodcock and Mark Graham, *The Gig Economy: A Critical Introduction* (London: Polity, 2019).

7. Gandini, "Digital Labour: An Empty Signifier?" 375.

8. David Stark and Ivana Pais, "Algorithmic Management in the Platform Economy," *Sociologica* 14, no. 3 (2020): 47–72.

9. Cyrille Schwellnus, Assaf Geva, Mathilde Pak, and Rafael Veiel, "Gig Economy Platforms: Boon or Bane?," OECD Economics Department Working Papers No. 1550

(2019): 6, accessed July 11, 2021, https://www.oecd-ilibrary.org/docserver/fdb0570b
-en.pdf?expires=1626013714&id=id&accname=guest&checksum=877B23ADA8FCE7
03DC98F0F863907AD2.

10. Masiero, "Digital Platform Workers under COVID-19: A Subaltern Perspective,"
6350.

11. Masiero, "Digital Platform Workers under COVID-19: A Subaltern Perspective,"
6350.

12. International Labor Organization (ILO), "World Employment Social Outlook
202: The Role of Digital Labour Platforms in Transforming the World of Work," 2022.
https://www.ilo.org/infostories/en-GB/Campaigns/WESO/World-Employment-Social
-Outlook-2021#introduction.

13. IMARC Group, "Online Food Delivery Services Global Market Report 2021:
COVID-19 Growth and Change to 2030," 2021, accessed August 19, 2021, https://
www.globenewswire.com/news-release/2021/05/04/2222008/28124/en/Global
-Online-Food-Delivery-Services-Market-Report-2021-Featuring-Market-Leaders
-takeaway-com-Doordash-Deliveroo-Uber-Eats-Zomato-Swiggy-Domino-s-Pizza
-Grubhub-foodpanda-and-Just-E.html.

14. Sam Pudwell, "How Deliveroo Is Using Big Data and Machine Learning to Power
Food Delivery," Silicon.co.uk, June 30, 2017, https://www.silicon.co.uk/data-storage
/bigdata/deliveroo-big-data-deliveries-216163.

15. Kathleen Griesbach, Adam Reich, Luke Elliott-Negri, and Ruth Milkman, "Algorithmic Control in Platform Food Delivery Work," *Socius* 5 (2019): 1–15, 2378023119870041.

16. Bonini and Gandini, "First Week Is Editorial, Second Week Is Algorithmic."
https://doi.org/10.1177/2056305119880006.

17. Francesca Sironi, "Io burattinaio dei rider vi racconto come controlliamo le
consegne e i fattorini," *L'Espresso*, December 20, 2018, https://espresso.repubblica.it
/attualita/2018/12/19/news/rider-controllo-foodora-1.329793/amp/.

18. Niels van Doorn and Julie Yujie Chen, "Odds Stacked against Workers: Datafied
Gamification on Chinese and American Food Delivery Platforms," *Socio-Economic
Review* 19, no. 4 (October 2021): 1345–1367. https://doi.org/10.1093/ser/mwab028.

19. According to Wikipedia, Honor of Kings (Chinese: 王者荣耀; pinyin: Wángzhě
Róngyào; lit. 'King's glory," unofficially translated as King of Glory, or alternatively
transliterated as "Wang Zhe Rong Yao," also known as "Honour of Kings"), is a
multiplayer online battle arena developed by TiMi Studio Group and published by
Tencent Games for the iOS and Android mobile platforms for the Chinese market.
Released in 2015, it has become the hottest MOBA game in mainland China. By
2017, *Honour of Kings* had over 80 million daily active players and 200 million
monthly active players, and was among the world's most popular and one of the
highest-grossing games of all time as well as the most downloaded app globally.
https://en.wikipedia.org/wiki/Honor_of_Kings (accessed August 18, 2021).

20. According to Lei, "Delivering Solidarity: Platform Architecture and Collective
Contention in China's Platform Economy," SPCs are full-time employees who work

for service stations (an actual physical station used to coordinate within a locality), while gig platform couriers (GPCs) can decide when they want to work and do not share a workplace.

21. Screenshot of a post published by a Chinese courier in a private WeChat group collected by Zizheng Yu, a Chinese PhD student supervised by Emiliano Treré at Cardiff University. He took part in our Algo Resistance research project. The screenshot was collected by permission of the author. See the appendix of this book for more details.

22. The interviews to the Spanish riders were collected by Francisco Javier Lòpez Ferràndez, a Spanish researcher from the Universitat Jaume I, Spain, that took part to our Algo Resistance research project. See the appendix of this book for more details.

23. On November 2, 2020, in Italy, the agreement signed on September 15, 2020 between the Unione Generale del Lavoro (UGL), a right-wing trade union, and Assodelivery, an Italian association that brings together the main companies in the Italian food delivery market (including Deliveroo, Glovo, Uber Eats, and Just Eat) came into force. The contract regulates the work of about 20,000 riders and recognizes their work as autonomous, to be carried out with "maximum independence and freedom." There is no mention of the possibility of this work being considered subordinate (and therefore of the rights that this would entail). The agreement foresees a minimum remuneration of 10 euros gross for each hour worked (this means that if a rider is online for three hours but receives only one order, which he delivers in ten minutes, he will be paid only for the ten minutes worked) and a supplementary allowance for night shifts, vacations, and work done in adverse weather conditions. The contract received a lot of criticism from left-wing trade unions and many riders. In the days following its implementation, protests erupted all over Italy and some turned into riots, as riders realized that their earnings were significantly worse than before.

24. Abel Guerra and Carlos d'Andréa, "Crossing the Algorithmic 'Red Sea': Brazilian Ubertubers' Ways of Knowing Surge Pricing," *Information, Communication & Society* (2022): 1–19.

25. Donatella della Porta, Riccardo Emilio Chesta, and Lorenzo Cini, "Mobilizing against the Odds: Solidarity in Action in the Platform Economy," *Berliner Journal für Soziologie* (2022): 1–29.

26. Ioulia Bessa, Simon Joyce, Denis Neumann, Mark Stuart, Vera Trappmann, and Charles Umney, "A Global Analysis of Worker Protest in Digital Labour Platforms," *ILO Working Paper* 70 (2022): 1–47.

27. See della Porta, Chesta, and Cini, "Mobilizing against the Odds. Solidarity in Action in the Platform Economy," 1.

28. Steve O'Hear, "Deliveroo Drivers Hold Protest in London over Possible Changes to the Way They Are Paid," *TechCrunch*, August 11, 2016, https://techcrunch.com /2016/08/11/deliveroo-drivers-hold-protest-in-london-over-possible-changes-to-the -way-they-are-paid/.

29. Jeremy Peck and Rachel Phillips, "The Platform Conjuncture," *Sociologica* 14, no. 3 (2020): 73–99. https://doi.org/10.6092/issn.1971-8853/11613.

30. See Romano Alquati, "Documenti sulla lotta di classe alla FIAT," *Quaderni rossi* 1 (1961): 198–244; Andrew Friedman, *Industry and Labour: Class Struggle at Work and Monopoly Capitalism* (New York: Macmillan International Higher Education, 1977); Michael Burawoy, *Manufacturing Consent: Changes in the Labor Process Under Monopoly Capitalism* (Chicago: University of Chicago Press, 1979); David Stark, "Class Struggle and the Transformation of the Labor Process," *Theory and Society* 9, no. 1 (1980): 89–130; Randy Hodson, "Worker Resistance: An Underdeveloped Concept in the Sociology of Work," *Economic and Industrial Democracy* 16, no. 1 (1995): 79–110.

31. Tiziana Terranova, "Free Labor: Producing Culture for the Digital Economy," *Social Text* 18, no. 2 (2000): 33–58; Ursula Huws, *Labor in the Global Digital Economy: The Cybertariat Comes of Age* (New York: New York University Press, 2014).

32. Scott, *Weapons of the Weak.*

33. Stellan Vinthagen and Anna Johansson, "Everyday Resistance: Exploration of a Concept and Its Theories," *Resistance Studies Magazine* 1 no. 1 (2013): 1–46.

34. Jenny Powers, "I'm a Part-Time Amazon Delivery Driver. Here's How We Cheat to Get around the Strict Rules and Constant Monitoring," *Business Insider*, April 6, 2021, https://www.businessinsider.com/amazon-delivery-driver-rules-tracking-app-pee -bottles-2021-4; See also Spencer Soper, "Amazon Drivers Are Hanging Smartphones in Trees to Get More Work," *Bloomberg*, September 1, 2020, https://www.bloomberg .com/news/articles/2020-09-01/amazon-drivers-are-hanging-smartphones-in-trees -to-get-more-work.

35. Alessandro Delfanti, *The Warehouse: Workers and Robots at Amazon* (London: Pluto Press, 2021).

36. Sam Adler-Bell, "Surviving Amazon," *Logic*, August 3, 2019, https://logicmag.io /bodies/surviving-amazon/.

37. Adler-Bell, "Surviving Amazon."

38. Adler-Bell, "Surviving Amazon."

39. Fabian Ferrari and Mark Graham, "Fissures in Algorithmic Power: Platforms, Code, and Contestation."

40. See the appendix of this book for more details on this aspect of our research.

41. Michel Foucault, *Technologies of the Self: A Seminar with Michel Foucault* (Amherst, MA: University of Massachusetts Press, 1988).

42. In this note, the courier keeps track of the distance traveled to complete the order, the costs incurred, and how much he has earned:

gross delivery price: 4.54 euro
distance from my location to the restaurant: 2.5 km
distance restaurant—customer: 9.9 km
distance customer—my location: 5.2 km
Total km driven: 17.6 km

Delivery time: about 40 minutes
total profit: 4.54 euro
Fuel cost: 1.50 euro
tax cost: 0.91 euro
net gain: 2.13 euro

net gain of 2.13 euro for 40 minutes of work

43. This was before Deliveroo decided to operate with free login in Italy, after November 2, 2021. On this date, Deliveroo introduced a new kind of contract and a free login for every worker. It means that every courier can work with Deliveroo whenever they want, without the need to book working hours in advance.

44. Ping Sun, "Your Order, Their Labor: An Exploration of Algorithms and Laboring on Food Delivery Platforms in China," *Chinese Journal of Communication* 12 no. 3 (2019): 308–323; Ping Sun and Julie Yujie Chen, "Platform Labour and Contingent Agency in China," *China Perspectives* 1 (2021): 19–27.

45. Francesco Bonifacio, "Easy, Rider? Pratiche, saperi e traiettorie di una professione emergente," PhD thesis, Università Cattolica del Sacro Cuore (2020/2021), 194.

46. Michel de Certeau, *The Practice of Everyday Life* (Berkeley: University of California Press, 1984).

47. Rida Qadri, "Delivery Platform Algorithms Don't Work without Drivers' Deep Local Knowledge," *Slate*, December 28, 2020, https://slate.com/technology/2020/12 /gojek-grab-indonesia-delivery-platforms-algorithms.html.

48. Edward Ongweso Jr. "Organized DoorDash Drivers' #DeclineNow Strategy Is Driving up Their Pay," *Vice*, February 19, 2021, https://www.vice.com/en/article /3anwdy/organized-doordash-drivers-declinenow-strategy-is-driving-up-their-pay.

49. Ongweso, "Organized DoorDash Drivers' #DeclineNow Strategy Is Driving up Their Pay."

50. Marco Briziarelli, "Spatial Politics in the Digital Realm: The Logistics/Precarity Dialectics and Deliveroo's Tertiary Space Struggles," *Cultural Studies* 33, no. 5 (2019): 823–840; Jack Shenker, "Strike 2.0: How Gig Economy Workers Are Using Tech to Fight Back," *The Guardian*, August 31, 2019, https://www.theguardian.com/books /2019/aug/31/the-new-resistance-how-gig-economy-workers-are-fighting-back.

51. Scott, *Weapons of the Weak.*

52. Gavin Mueller, *Breaking Things at Work* (London: Verso, 2021), 16.

53. Winner, "Do Artifacts Have Politics?"

54. Antonio Aloisi and Valerio De Stefano, *Your Boss Is an Algorithm: Artificial Intelligence, Platform Work and Labour* (London: Bloomsbury, 2022).

55. Kylie Jarrett, *Digital Labour* (London: Polity, 2022).

56. Jenny L. Davis, *How Artifacts Afford: The Power and Politics of Everyday Things* (Cambridge, MA: MIT Press, 2020).

57. Rida Qadri, "Delivery Drivers Are Using Grey Market Apps to Make Their Jobs Suck Less," *Vice*, April 27, 2021, https://www.vice.com/en/article/7kvpng/delivery -drivers-are-using-grey-market-apps-to-make-their-jobs-suck-less.

58. Varsha Bansal, "Meet the Most Powerful Uber Driver in India," *Rest of World*, January 4, 2023, https://restofworld.org/2023/india-powerful-uber-driver/.

59. Godofredo Ramizo Jr., "Platform Playbook: A Typology of Consumer Strategies against Algorithmic Control in Digital Platforms," *Information, Communication & Society* (2021):1–16. https://doi.org/10.1080/1369118X.2021.1897151.

60. Ludmila C. Abilio, Rafael Grohmann, and Henrique C. Weiss, "Struggles of Delivery Workers in Brazil: Working Conditions and Collective Organization during the Pandemic," *Journal of Labor and Society* 24, no. 4 (2021): 598–616.

61. Michael Maffie, "The Role of Digital Communities in Organizing Gig Workers," *Industrial Relations* 59, no. 1 (2020): 123–149.

62. Woodcock, *The Fight Against Platform Capitalism*, 72.

63. Soriano and Cabañes, "Entrepreneurial Solidarities: Social Media Collectives and Filipino Digital Platform Workers."

64. Taina Bucher, "The Algorithmic Imaginary: Exploring the Ordinary Affects of Facebook Algorithms," *Information, Communication & Society* 20, no.1 (2017): 30–44.

65. Michael A. DeVito, Jeremy Birnholtz, Jeffrey T. Hancock, Megan French, and Sunny Liu, "How People Form Folk Theories of Social Media Feeds and What It Means for How We Study Self-presentation," in *Proceedings of the 2018 CHI conference on human factors in computing systems* (April 2018): 1–12, https://dl.acm.org/doi/pdf /10.1145/3173574.3173694.

66. Anne-Britt Gran, Peter Booth, and Taina Bucher, "To Be or Not to Be Algorithm Aware: A Question of a New Digital Divide?" *Information, Communication & Society* 24, no. 12 (2021): 1779–1796. https://doi.org//10.1080/1369118X.2020.1736124.

67. James C. Scott, *Domination and the Arts of Resistance: Hidden Transcripts* (New Haven, CT: Yale University Press, 1990); Mohammad A. Anwar and Mark Graham, "Hidden Transcripts of the Gig Economy: Labour Agency and the New Art of Resistance among African Gig Workers," *Environment and Planning A: Economy and Space* 52, no. 7 (2020): 1269–1291.

68. Scott, *Domination and the Arts of Resistance*.

69. Scott, *Domination and the Arts of Resistance*, 2.

70. Scott, *Domination and the Arts of Resistance*, 4, quoted in Anwar and Graham, "Hidden Transcripts of the Gig Economy: Labour Agency and the New Art of Resistance among African Gig Workers," 1272.

71. Woodcock, *The Fight Against Platform Capitalism*, 72.

72. Nishant Kauntia, "How Swiggy Threatened to "Suspend" Protesting Delhi Workers after Second Pay Cut in Seven Months," *The Caravan*, August 27, 2020, https:// caravanmagazine.in/news/swiggy-pay-cut-delhi-worker-delivery-protests-aigwu.

73. These data come from the "Collective Action in Tech" website, by a volunteer-run nonprofit organization that "document those collective actions and has evolved into a platform for workers to tell their stories, share resources, and theorize the tech workers' movement together. See: https://data.collectiveaction.tech/?query=swiggy.

74. Arianna Tassinari and Vincenzo Maccarrone, "Couriers on the Storm: Workplace Solidarity among Gig Economy Couriers in Italy and the UK," *Work, Employment and Society* 34, no. 1 (2020): 35–54. https://doi.org/10.1177/0950017019862954.

75. Zello is an instant communication app to exchange short audio messages. It simulates traditional two-way radios with a walkie-talkie style of communication.

76. Kimberly Mutandiro, "Deadly Robberies Force Bolt Drivers to Create Self-defense Groups in South Africa," *Rest of World*, October 6, 2022, https://restofworld.org/2022/deadly-robberies-force-bolt-drivers-to-create-self-defense-groups-in-south-africa/.

77. "Knights" means "Couriers" here. According to the interviewees from the KL, the word "knight" has a higher status than "courier," and Chinese couriers prefer use that word to refer to themselves.

78. Soriano and Cabañes, "Entrepreneurial Solidarities: Social Media Collectives and Filipino Digital Platform Workers."

79. Soriano and Cabañes, "Entrepreneurial Solidarities: Social Media Collectives and Filipino Digital Platform Workers," 9.

80. Guest and Zhou, "The Gig Workers Index: Mixed Emotions, Dim Prospects."

81. The existence of at least two different moral economies built around different labor ideologies is also supported by other studies, such as that of the Mexican sociologist Mariana Manriquez among Uber drivers in Monterrey, Mexico, and that of Krishnan Vasudevan and Ngai Keung Chan. Both studies analyzed how gig workers collaboratively develop strategies and "work games" to protect their autonomy and maximize their earnings in the face of gamified algorithmic management. As Manriquez argued, modes of play contribute to understanding how a driver's strategy either sustains or diverges from the "firm's entrepreneurial ideology." Building upon this premise, Vasudevan and Chan reveal two distinctive player modes, "grinding" and "oppositional" play, which articulate consent to gamification and resistance through the creation of work games, respectively. See Mariana Manriquez, "Work-Games in the Gig-Economy: A Case Study of Uber Drivers in the City of Monterrey, Mexico," *Research in the Sociology of Work* 33 (2019): 165–188; Vasudevan and Chan, "Gamification and Work Games: Examining Consent and Resistance among Uber Drivers."

82. The term "Stakhanovite" (стахановец) originated in the Soviet Union and referred to workers who modeled themselves after Alexei Stakhanov. These workers took pride in their ability to produce more than was required, by working harder and more efficiently, thus strengthening the socialist state. In common parlance, it designates a tireless and productive worker. Vladimir Shlapentokh, "The Stakhanovite Movement: Changing Perceptions over Fifty Years," *Journal of Contemporary History*, 23 no. 2 (1988): 259–276.

83. This desire for freedom, combined with precarious and low-paid work, reminds us that the reasons why people work go beyond simple economics. The search for personal freedom in the work of couriers reminds us closely of the forms of freedom that the Matsutake mushroom hunters described by the anthropologist Anne Lowenthal Tsing pursued in their precarious work in the forests of Oregon. See Anna Lowenhaupt Tsing, *The Mushroom at the End of the World* (Princeton, NJ: Princeton University Press, 2015).

84. Vasudevan and Chan, "Gamification and Work Games: Examining Consent and Resistance among Uber Drivers," 876.

85. Charlesworth and Randall, "Morals, Markets and the English Crowd in 1766," 201.

86. See Robin Food, https://www.robinfoodfirenze.it/ (accessed September 30, 2021).

87. Coop Cycle is a cooperative that provides its members with a technological platform that they could not otherwise afford on their own. The software developed by the co-op is also released as a "digital common" and protected by an ad hoc license, Coopyleft, which allows access to the code only to nonprofit cooperative entities. At the end of 2021, Coop Cycle counted seventy members around the world, mostly concentrated in Europe. See https://coopcycle.org/en/ (accessed October 4, 2021).

88. Zigor Aldama, "'Couriers' vascos se unen en cooperativas para competir con Glovo y Deliveroo," *El Diario Vasco*, January 31, 2021, https://www.diariovasco.com/economia/couriers-vascos-unen-20210131004832-ntvo.html.

89. Trebor Scholz, *Platform Cooperativism. Challenging the Corporate Sharing Economy* (New York: Rosa Luxemburg Foundation, 2016).

90. Megan Carnegie, "Worker-Owned Apps Are Redefining the Sharing Economy," *Wired*, June 30, 2022. https://www.wired.com/story/gig-economy-worker-owned-apps/.

91. Carnegie, "Worker-Owned Apps Are Redefining the Sharing Economy."

92. Carnegie, "Worker-Owned Apps Are Redefining the Sharing Economy."

CHAPTER 4

1. Nieborg and Poell, "The Platformization of Cultural Production."

2. Thomas Poell, David B. Nieborg, and José van Dijck, "Platformisation," *Internet Policy Review* 8, no. 4 (2019): 1–13.

3. Brooke E. Duffy, Thomas Poell, and David B. Nieborg, "Platform Practices in the Cultural Industries: Creativity, Labor, and Citizenship," *Social Media + Society* 5, no. 4 (October 2019): 1. https://doi.org/10.1177/2056305119879672.

4. Thomas Poell, David B. Nieborg, and Brooke E. Duffy, *Platforms and Cultural Production* (London: Polity, 2022).

5. Poell, Nieborg, and Duffy, *Platforms and Cultural Production*, 6.

6. Rasmus Kleis Nielsen and Sarah Anne Ganter, *The Power of Platforms: Shaping Media and Society* (Oxford: Oxford University Press, 2022).

7. This case history is described in Gillespie, *Custodians of the Internet*.

8. Nieborg and Poell, "The Platformization of Cultural Production," 4281.

9. Poell, Nieborg, and Duffy, *Platforms and Cultural Production*, 81 (emphasis in original).

10. Gillespie, "Algorithmically Recognizable"; Jeremy W. Morris, "Music Platforms and the Optimization of Culture," *Social Media + Society* 6, no. 3 (July 2020); https://doi.org/10.1177/2056305120940690; Jeremy W. Morris, Robert Prey, and David B. Nieborg, "Engineering Culture: Logics of Optimization in Music, Games, and Apps," *Review of Communication* 21, no. 2 (2021): 161–175.

11. Liz Pelly, "Streambait Pop," *The Baffler*, December 11, 2018, https://thebaffler .com/downstream/streambait-pop-pelly, quoted in Morris, "Music Platforms and the Optimization of Culture," 1.

12. Morris, "Music Platforms and the Optimization of Culture," 1.

13. Nieborg and Poell, "The Platformization of Cultural Production," 4282.

14. Bonini and Gandini, "First Week Is Editorial, Second Week Is Algorithmic."

15. Robert Prey, "Nothing Personal: Algorithmic Individuation on Music Streaming Platforms," *Media, Culture & Society* 40 no. 7 (2018): 1086–1100.

16. Zoe Glatt, "We're All Told not to Put Our Eggs in One Basket: Uncertainty, Precarity and Cross-platform Labor in the Online Video Influencer Industry," *International Journal of Communication*, 16 (2021): 1–19.

17. Rosalind Gill, "Life Is a Pitch: Managing the Self in New Media Work," in *Managing Media Work*, ed. Mark Deuze (London: SAGE, 2011), 249–262; Rosalind Gill and Andy Pratt, "In the Social Factory? Immaterial Labour, Precariousness and Cultural Work," *Theory, Culture & Society* 25, no. 7–8 (2008): 1–30; David Hesmondhalgh and Sarah Baker, *Creative Labour: Media Work in Three Cultural Industries* (London: Routledge, 2013); Angela McRobbie, *Be Creative: Making a Living in the New Culture Industries* (London: Polity, 2016); Brooke E. Duffy, *(Not) Getting Paid to Do What You Love* (New Haven, CT: Yale University Press, 2018).

18. Honoré De Balzac, *Lost Illusions* (London: Penguin Classics, 1976).

19. David Hesmondhalgh, *The Cultural Industries* (London: SAGE, 2018).

20. Brooke E. Duffy, Annika Pinch, Shruti Sannon, and Megan Sawey, "The Nested Precarities of Creative Labor on Social Media," *Social Media + Society* 7, no. 2 (April 2021): 2. https://doi.org/10.1177/20563051211021368.

21. Duffy, Pinch, Sannon, and Sawey, "The Nested Precarities of Creative Labor on Social Media," 1.

22. Abidin defines visibility labor as "the work individuals do when they self-posture and curate their self-presentations so as to be noticeable and positively prominent among prospective employers, clients, the press or followers and fans. . . . . Visibility labour is concerned with analogue affective labour ordinary users perform to be noticed by prolific elite users" (2016, 90). Chrystal Abidin, "Visibility Labour:

Engaging with Influencers' Fashion Brands and# OOTD Advertorial Campaigns on Instagram," *Media International Australia* 161, no. 1 (2016): 86–100.

23. Kelley Cotter, "Playing the Visibility Game: How Digital Influencers and Algorithms Negotiate Influence on Instagram," *New Media & Society* 21, no. 4 (2019): 895–913.

24. See Robert Prey, "Nothing Personal: Algorithmic Individuation on Music Streaming Platforms," *Media, Culture & Society* 40, no. 7 (2018): 1086–1100.

25. Duffy, Pinch, Sannon, and Sawey, "The Nested Precarities of Creative Labor on Social Media," 4.

26. Duffy, Pinch, Sannon, and Sawey, "The Nested Precarities of Creative Labor on Social Media," 4.

27. Taina Bucher, "Want to Be on the Top? Algorithmic Power and the Threat of Invisibility on Facebook," *New Media & Society* 14, no. 7 (2012): 1164–1180.

28. Brooke E. Duffy, "Algorithmic Precarity in Cultural Work," *Communication and the Public* 5, no. 3–4 (2020): 103–107.

29. Sophie Bishop, "Algorithmic Experts: Selling Algorithmic Lore on YouTube," *Social Media+Society* 6, no. 1 (January 2020). https://doi.org/10.1177/2056305119897323; quoted in Duffy, Pinch, Sannon, and Sawey, "The Nested Precarities of Creative Labor on Social Media," 9.

30. David B. Nieborg, Thomas Poell, and Brooke E. Duffy, "Analyzing Platform Power in the Cultural Industries," *AoIR Selected Papers of Internet Research* (September 2021). https://spir.aoir.org/ojs/index.php/spir/article/view/12219.

31. Wang Xiuying, "The Word from Wuhan," *London Review of Books* 42, no. 5 (March 2020). https://www.lrb.co.uk/the-paper/v42/n05/wang-xiuying/the-word-from-wuhan.

32. Bo Zhao and Qinying Chen, "Location Spoofing in a Location-Based Game: A Case Study of Pokémon Go," *International Cartographic Conference* (2017): 21–32. https://doi.org/10.1007/978-3-319-57336-6_2.

33. Matt Brett, "Digital EPO: Smash your Strava Times . . . by Cheating," *Road*, June 4, 2013, https://road.cc/content/news/84868-digital-epo-smash-your-strava-times %E2%80%A6-cheating; Olga Khazan, "How to Fake Your Workout," *The Atlantic*, September 28, 2015, https://www.theatlantic.com/health/archive/2015/09/unfit-bits /407644/.

34. Susanna Bonelli, "Young People and Love: How Dating Apps Interfere with Affective Relationship Formation," MA thesis, University of Siena, 2020.

35. Jeremy W. Morris, "Infrastructures of Discovery: Examining Podcast Ratings and Rankings," *Cultural Studies* 35, no. 4–5 (2021): 728–749. https://doi.org/10.1080 /09502386.2021.1895246.

36. Amy X. Wang, "A Bulgarian Scheme Scammed Spotify for $1 Million—Without Breaking a Single Law," *Quartz*, February 22, 2018, https://qz.com/1212330/a -bulgarian-scheme-scammed-spotify-for-1-million-without-breaking-a-single-law

/; Michael Byrne, "The Silent Spotify Album 'Sleepify' Made $20,000 in Royalties," *Vice*, July 26, 2014. https://www.vice.com/en/article/mgb394/the-silent-spotify-album-sleepify-made-20000-in-royalties; Blake Montgomery, "Fans Are Spoofing Spotify with "Fake Plays" and That's a Problem for Music Charts," *BuzzFeed News*, September 13, 2018, https://www.buzzfeednews.com/article/blakemontgomery/spotify-billboard-charts.

37. Eamonn Forde, "Ex-Stream Measures: Spotify Clamps down on 'Artificial Plays,'" *Forbes*, October 1, 2021. https://www.forbes.com/sites/eamonnforde/2021/10/01/ex-stream-measures-spotify-clamps-down-on-artificial-plays/.

38. Taylor Lorenz, Kellen Browning, and Sheera Frenkel, "Tik-Tok Teens and K-Pop Stans Say They Sank Trump Rally," *New York Times*, June 21, 2020, https://www.nytimes.com/2020/06/21/style/tiktok-trump-rally-tulsa.html.

39. Blake Montgomery, "Fans are Spoofing Spotify with "Fake Plays," and That's a Problem for Music Charts," *BuzzFeed News*, September 13, 2018, https://www.buzzfeednews.com/article/blakemontgomery/spotify-billboard-charts.

40. Haeryun Kang, "Inside Sajaegi, K-Pop's Open Secret," *NPR*, February 21, 2020, https://www.npr.org/2020/02/21/808049441/inside-sajaegi-k-pops-open-secret.

41. Qian Zhang and Keith Negus, "East Asian Pop Music Idol Production and the Emergence of Data Fandom in China," *International Journal of Cultural Studies* 23, no. 4 (2020): 493–511.

42. Zhang and Negus, "East Asian Pop music idol production and the emergence of data fandom in China," 494.

43. Kang, "Inside Sajaegi, K-Pop's Open Secret."

44. Kang, "Inside Sajaegi, K-Pop's Open Secret."

45. Glatt, *"We're All Told Not to Put Our Eggs in One Basket,"* 7.

46. See the appendix of the book for research details.

47. Victoria O'Meara, "Weapons of the Chic: Instagram Influencer Engagement Pods as Practices of Resistance to Instagram Platform Labor," *Social Media + Society* 5, no. 4 (October 2019): 1–11. https://doi.org/10.1177/2056305119879671.

48. Instagram 2016, quoted in Martina Mahnke Skrubbeltrang, Josefine Grunnet, and Nicolai Traasdahl Tarp, "# RIPINSTAGRAM: Examining User's Counter-narratives Opposing the Introduction of Algorithmic Personalization on Instagram," *First Monday* (2017), https://firstmonday.org/ojs/index.php/fm/article/download/7574/6095.

49. Skrubbeltrang, Grunnet, and Tarp, "# RIPINSTAGRAM: Examining User's Counter-narratives Opposing the Introduction of Algorithmic Personalization on Instagram."

50. Sarah Heard, "Keep *Instagram* Chronological," *Change.org,* March 2016, https://www.change.org/p/keep-instagram-chronological.

51. O'Meara, "Weapons of the Chic," 6.

52. See Reddit conversations about this practice: https://www.reddit.com/r/socialmedia/comments/igk9fp/tiktok_engagement_pod/.

53. Motahare Eslami, Kerrie Karahalios, Christian Sandvig, et al., "First I 'Like' It, then I Hide It: Folk Theories of Social Feeds," in *Proceedings of the 2016 CHI Conference on Human Factors in Computing Systems* (May 2016): 2371–2382. See also Michael A. DeVito, Darren Gergle, and Jeremy Birnholtz, "Algorithms Ruin Everything" # RIPTwitter, Folk Theories, and Resistance to Algorithmic Change in Social Media," in *Proceedings of the 2017 CHI Conference on Human Factors in Computing Systems* (May 2017): 3163–3174.

54. Crystal Abidin, *Internet Celebrity: Understanding Fame Online* (Bingley, UK: Emerald Publishing Ltd., 2018).

55. O'Meara, "Weapons of the Chic."

56. Microinfluencers are individuals who tend to have followings from 10,000 to 50,000, while nanoinfluencers have fewer than 10,000 followers. See O'Meara, "Weapons of the Chic."

57. *"Sotto torchio"* is an Italian expression that means "under pressure." *Torchio* literally means "the press," and it is normally used to refer to people under investigation by the police.

58. Scott, *The Moral Economy of the Peasant*, 1. Scott takes the citation from Richard Henry Tawney, *Land and Labour in China* (London: George Allen and Unwin Ltd., 1932), 77.

59. On the precarious condition of content creators in the platformized cultural industries, see Brooke E. Duffy and Elizabeth Wissinger, "Mythologies of Creative Work in the Social Media Age: Fun, Free, and 'Just Being Me,'" *International Journal of Communication* 11, no. 20 (2017). https://ijoc.org/index.php/ijoc/article/view/7322; Duffy, *(Not) Getting Paid to Do What You Love*; Brooke E. Duffy, "The Romance of Work: Gender and Aspirational Labour in the Digital Culture Industries," *International Journal of Cultural Studies* 19, no. 4 (2016): 441–457.

60. Hannah Gelbart, Mamdouh Akbiek, and Ziad Al-Qattan, "Tik Tok Profits from Livestreams of Families Begging," *BBC News*, October 12, 2022. https://www.bbc.com/news/world-63213567.

61. Duffy, "Algorithmic Precarity in Cultural Work," 103.

62. Glatt, *"We're All Told Not to Put Our Eggs in One Basket,"* 1.

63. See, for example, the "rules" set by the admin of this TikTok engagement group: "Hi! My name is Shayla Terkalas and I am a Social Media Coach+Tik Tok Engagement Specialist. I have created this group to help influencers and business owners grow their audience and engage with each other on Tik Tok.

The group is open to all: lifestyle, fitness, fashion, business, influencer and more.
In order to be part of the group you MUST follow the rules.
Admins will be monitoring threads to make sure no robots or fake accounts are cluttering the
space.

RULES
- You must be 16 years or older
- NO spam is allowed & NO affiliate links!

- Be friendly & kind to one another. If you are not, you will be removed immediately.
- YOU MUST FOLLOW ADMIN TIKTOK ACCOUNT IN ORDER TO REMAIN PART OF THE
  GROUP:

TikTok Handle: HowToGrowOnTikTok

This is the perfect place to ask your questions, recommendations, find collaboration opportunities and make new friends!

Thanks,

Shayla Terkalas. Available at: https://www.facebook.com/groups/260837715325961 (accessed October 28, 2021).

64. Here is our translation of the rules shown in figure 4.3:

1. Send the link to your Instagram post, just the link, without anything else. to copy the link click on the 3 dots at the top right of the photo on the group
2. Comment and like the 5 messages above your newly posted link. you can use the command/list to receive the list of the 5 links to like and comment (both mandatory)
3. The rules above must be respected, no exceptions! Both likes and comments are mandatory.
4. Those who do not follow the rules, after 5 warnings, will be deleted immediately.

65. "Hello everyone, this group was born from the idea of growing together, exchanging follow and like, invite all your friends so that this community grows and help us all together. good permanence to all."

66. Marcel Mauss, *The Gift: The Form and Reason for Exchange in Archaic Societies* (London: Cohen & West, 1954).

67. Scott, *Weapons of the Weak*, 262.

68. Carolin Gerlitz and Anne Helmond, "The Like Economy: Social Buttons and the Data-Intensive Web," *New Media & Society* 15, no. 8 (2013): 1348–1365.

69. Woodcock and Graham, *The Gig Economy*.

70. IG Metall, "'FairTube': IG Metall and Youtubers Union Launch Joint Initiative," *IG Metall*, July 22, 2019, https://www.igmetall.de/download/20190722 _Presseinformation_Joint_Venture_EN_35b71bba37474446ae6e3618b43d061c6e0ed 9af.pdf. See also Valentin Niebler, "'YouTubers Unite': Collective Action by YouTube Content Creators," *Transfer: European Review of Labour and Research* 26, no. 2 (2020): 223–227.

71. Amelia Tait, "'Influencers Are Being Taken Advantage Of': The Social Media Stars Turning to Unions," *The Guardian*, October 10, 2020, https://www.theguardian .com/media/2020/oct/10/influencers-are-being-taken-advantage-of-the-social-media -stars-turning-to-unions.

CHAPTER 5

1. Guiomar Rovira Sancho, "Tecnopolítica para la emancipación y para la guerra: acción colectiva y contrainsurgencia," *IC: Revista Científica de Información y Comunicación* 16 (2019): 39–83.

2. Evelyn Ruppert, Engin Isin, and Didier Bigo, "Data Politics," *Big Data & Society* 4, no. 2 (July 2017): 1–7.

3. Ruppert, Isin, and Bigo, "Data Politics," 1.

4. Ruppert, Isin, and Bigo, "Data Politics," 2.

5. Charles Tilly and Sidney G. Tarrow, *Contentious Politics*. 2nd ed. (New York: Oxford University Press, 2015).

6. Davide Beraldo and Stefania Milan, "From Data Politics to the Contentious Politics of Data," *Big Data & Society* 6 no. 2 (2019): 1–11.

7. Beraldo and Milan, "From Data Politics to the Contentious Politics of Data," 2.

8. Beraldo and Milan, "From Data Politics to the Contentious Politics of Data," 2.

9. Samuel C. Woolley, *The Reality Game: A Gripping Investigation into Deepfake Videos, the Next Wave of Fake News and What It Means for Democracy* (London: Hachette UK, 2020).

10. Ico Maly, "Algorithmic Populism and Algorithmic Activism," *Diggit Magazine*, October 28, 2018, https://www.diggitmagazine.com/articles/algorithmic-populism -activism; Ico Maly, "New Right Metapolitics and the Algorithmic Activism of Schild & Vrienden," *Social Media + Society* 5, no. 2 (April–June 2019): 1–15.

11. Emiliano Treré, *Hybrid Media Activism: Ecologies, Imaginaries, Algorithms* (London: Routledge, 2018); Emiliano Treré, "From Digital Activism to Algorithmic Resistance," in *Routledge Companion to Media and Activism*, ed. Graham Meikle (New York: Routledge, 2018), 367–375.

12. Maly "Algorithmic Populism and Algorithmic Activism."

13. See, for example, Peter Pomerantsev, *This Is not Propaganda* (London: Faber & Faber, 2019); Marco T. Bastos and Dan Mercea, "The Brexit Botnet and User-generated Hyperpartisan News," *Social Science Computer Review* 37, no. 1: 38–54; Yochai Benkler, Robert Faris, and Hal Roberts, *Network Propaganda: Manipulation, Disinformation, and Radicalization in American Politics* (Oxford: Oxford University Press, 2018); on the COVID-19 infodemic, see Matteo Cinelli, Walter Quattrociocchi, Alessandro Galeazzi, et al., "The COVID-19 Social Media Infodemic," *Nature Scientific Reports* 10, no. 1 (2020): 1–10. https://www.nature.com/articles/s41598-020-73510-5.

14. Claire Wardle, "Understanding Information Disorder," *First Draft* (October 2019), 6. https://firstdraftnews.org/wpcontent/uploads/2019/10/Information_Disorder_Digital _AW.pdf?x76701.

15. Thorsten Quandt, Lena Frischlich, Svenja Boberg, and Tim Schatto-Eckrodt, "Fake News," in *International Encyclopedia of Journalism Studies*, eds. Tim P. Vos and Folker Hanusch (Malden, MA: Wiley-Blackwell, 2019), 1–6.

16. Matthew d'Ancona, *Post-Truth: The New War on Truth and How to Fight Back* (London: Random House, 2017).

17. Quandt, Frischlich, Boberg, and Schatto-Eckrodt, "Fake News," 3.

18. Claire Wardle and Hossein Derakhshan, "Information Disorder: Toward an Interdisciplinary Framework for Research and Policymaking," Council of Europe Report (2017).

19. Thorsten Quandt, "Dark Participation," *Media and communication* 6 no. 4 (2018): 36–48.

20. Ico Maly, "Algorithmic Populism and the Datafication and Gamification of the People by Flemish Interest in Belgium," *Trabalhos em Linguística Aplicada* 59 (2020): 444–468.

21. Andrew Chadwick, *The Hybrid Media System: Politics and Power* (Oxford: Oxford University Press, 2017).

22. Paolo Gerbaudo, "Social Media and Populism: An Elective Affinity?" *Media, Culture & Society* 40 no. 5 (2018): 745–753.

23. Samuel C. Woolley and Philip N. Howard, "Automation, Algorithms, and Politics| Political Communication, Computational Propaganda, and Autonomous Agents—Introduction," *International Journal of Communication*, 10 (2016), 4882–4890; Samuel C. Woolley and Philip N. Howard, eds., *Computational Propaganda: Political Parties, Politicians, and Political Manipulation on Social Media* (Oxford: Oxford University Press, 2018); Woolley, *The Reality Game*; Samuel C. Woolley, "Bots and Computational Propaganda: Automation for Communication and Control," in *Social Media and Democracy: The State of the Field, Prospects for Reform*, eds. Nathaniel Persily and Joshua A. Tucker (Cambridge: Cambridge University Press, 2020), 89–110.

24. Woolley and Howard, "Automation, Algorithms, and Politics| Political Communication, Computational Propaganda, and Autonomous Agents—Introduction."

25. According to the *Merriam-Webster Dictionary*, the term "astroturfing" indicates any organized activity that is intended to create a false impression of a widespread, spontaneously arising, grassroots movement in support of or in opposition to something (such as a political policy), but that is in reality initiated and controlled by a concealed group or organization (such as a corporation). Classic astroturfing is the practice of disguising an orchestrated campaign as a spontaneous upwelling of public opinion.

26. Jacob Ratkiewicz, Michael Conover, Mark Meiss, et al., "Truthy: Mapping the Spread of Astroturf in Microblog Streams," *Proceedings of the 20th International Conference Companion on World Wide Web*, Hyderabad, India: ACM (2011): 249–252. Also see Woolley and Howard, "Automation, Algorithms, and Politics| Political Communication, Computational Propaganda, and Autonomous Agents—Introduction"; Woolley, "Bots and Computational Propaganda."

27. Norah Abokhodair, Daisy Yoo, and David W. McDonald, "Dissecting a Social Botnet: Growth, Content and Influence in Twitter," *Proceedings of the 18th ACM Conference on Computer Supported Cooperative Work & Social Computing* (2015); Emiliano Treré, "The Dark Side of Digital Politics: Understanding the Algorithmic Manufacturing of Consent and the Hindering of Online Dissidence," *IDS Bulletin* 47, no. 1 (2016). https://bulletin.ids.ac.uk/index.php/idsbo/article/view/41/html; Pablo Suárez-Serrato, Margaret E. Roberts, Clayton A. Davis, and Filippo Menczer, "On the Influence of Social Bots in Online Protests," *International Conference on Social Informatics* (Cham, Switzerland: Springer Cham, 2016).

28. Paula Chakravartty and Srirupa Roy, "Mediatized Populisms: Inter-Asian Lineages," *International Journal of Communication* 11 (2017): 4073–4092.

29. Chakravartty and Roy, "Mediatized Populisms: Inter-Asian Lineages," 4076.

30. Gillian Bolsover and Philip N. Howard, "Computational Propaganda and Political Big Data: Moving toward a More Critical Research Agenda," *Big Data* 5, no. 4 (2017): 273–276; Woolley, and Howard, eds., *Computational Propaganda*.

31. Benkler, Faris, and Roberts, *Network Propaganda*, 22; see also Silvio Waisbord, "Truth is What Happens to News: on Journalism, Fake News, and Post-truth," *Journalism Studies* 19, no. 13 (2018): 1866–187.

32. Jonathan C. Ong and Jason Vincent A. Cabañes, "Architects of Networked Disinformation: Behind the Scenes of Troll Accounts and Fake News Production in the Philippines," Newton Tech4Dev (2018), http://newtontechfordev.com/wp-content /uploads/2018/02/ARCHITECTS-OF-NETWORKED-DISINFORMATION-FULL-REPORT .pdf.

33. Treré, "The Dark Side of Digital Politics."

34. Stanley Cohen, *Folk Devils and Moral Panics* (London: MacGibbon and Kee, 1972).

35. Jason Vincent A. Cabañes, "Digital Disinformation and the Imaginative Dimension of Communication," *Journalism & Mass Communication Quarterly* 97, no. 2 (2020): 435–452.

36. Jonathan C. Ong and Jason Vincent A. Cabañes, "When Disinformation Studies Meets Production Studies: Social Identities and Moral Justifications in the Political Trolling Industry," *International Journal of Communication* 13 (2019): 5771–5790.

37. Ong and Cabañes, "When Disinformation Studies Meets Production Studies: Social Identities and Moral Justifications in the Political Trolling Industry," 5776.

38. Stefania Milan, "From Social Movements to Cloud Protesting: The Evolution of Collective Identity," *Information, Communication & Society* 18, no. 8 (2015): 887–900.

39. Treré, *Hybrid Media Activism: Ecologies, Imaginaries, Algorithms*; Beraldo and Milan, "From Data Politics to the Contentious Politics of Data."

40. Ulrich Dolata, "Social Movements and the Internet: The Sociotechnical Constitution of Collective Action," Research Contributions to Organizational Sociology and Innovation Studies, SOI Discussion Paper 02 (2017). https://www.sowi .uni-stuttgart.de/dokumente/forschung/soi/soi_2017_2_Dolata.Internet.and.Social .Movements.pdf.

41. Evelyn Ruppert, *Sociotechnical Imaginaries of Different Data Futures: An Experiment in Citizen Data* (Rotterdam: Erasmus University Press, 2018).

42. Helen Kennedy, "Living with Data: Aligning Data Studies and Data Activism through a Focus on Everyday Experiences of Datafication," *Krisis: Journal for Contemporary Philosophy* 1 (2018): 18–30.

43. Stefania Milan, "Political Agency, Digital Traces, and Bottom-up Data Practices," *International Journal of Communication* 12 (2018): 507–525.

44. Jonathan Gray, "Three Aspects of Data Worlds," *Krisis: Journal for Contemporary Philosophy* 1 (2018): 5–17.

45. Couldry and Powell, "Big Data from the Bottom Up."

46. Stefania Milan, "Data Activism as the New Frontier of Media Activism," in *Media Activism in the Digital Age*, eds. Goubin Yang and Viktor Pickard (New York: Routledge, 2017), 151–163.

47. Kennedy, "Living with Data: Aligning Data Studies and Data Activism Through a Focus on Everyday Experiences of Datafication."

48. Maly "Algorithmic Populism and Algorithmic Activism"; Maly, "New Right Metapolitics and the Algorithmic Activism of Schild & Vrienden."

49. Maly, "New Right Metapolitics and the Algorithmic Activism of Schild & Vrienden," 1.

50. The management scholars Katherine Kellogg, Melissa Valentine, and Angele Christin use a similar term, "algoactivism," to make sense of the individual and collective tactics of gig workers against algorithmic control (see chapter 3): Katherine C. Kellogg, Melissa A. Valentine, and Angele Christin, "Algorithms at Work: The New Contested Terrain of Control," *Academy of Management Annals* 14, no. 1 (2020): 366–410.

51. Beraldo and Milan, "From Data Politics to the Contentious Politics of Data," 6 (emphasis added).

52. Beraldo and Milan, "From Data Politics to the Contentious Politics of Data," 6.

53. Charles Tilly, *Contentious Performances* (Cambridge: Cambridge University Press, 2008).

54. Beraldo and Milan, "From Data Politics to the Contentious Politics of Data," 6.

55. Eubanks, *Automating Inequality: How High-Tech Tools Profile, Police, and Punish the Poor*.

56. Noble, *Algorithms of Oppression: How Search Engines Reinforce Racism*.

57. Benjamin, *Race after Technology: Abolitionist Tools for the New Jim Code*.

58. Costanza-Chock, *Design Justice: Community-led Practices to Build the Worlds We Need*.

59. Dan McQuillan, *Resisting AI: An Anti-fascist Approach to Artificial Intelligence* (Bristol: Bristol University Press, 2022).

60. See, for example, Algorithm Watch, the Algorithmic Justice League, the Mozilla Foundation, the Ada Lovelace Institute, Data for Black Lives, the DAIR Institute, the AI on the Ground Initiative, the BigDataSur Initiative, and The Data Justice Lab.

61. Charles Tilly, *The Contentious French: Four Centuries of Popular Struggle* (Cambridge, MA: Belknap Press, 1986), 2.

62. Charles Tilly, *Regimes and Repertoires* (Chicago: University of Chicago Press, 2010), 35.

63. Tilly, *Contentious Performances*.

64. Brett Rolfe, "Building an Electronic Repertoire of Contention," *Social Movement Studies* 4, no. 1 (2005): 65–74.

65. Yana Breindl and François Briatte, "Digital Network Repertoires and the Contentious Politics of Digital Copyright in France and the European Union," *Policy and Internet*, no 5 (2013): 27–55.

66. Jun Liu, "Technology for Activism: Toward a Relational Framework." *Computer Supported Cooperative Work (CSCW)* 30, no. 5 (2021): 627–650

67. Vasilis Galis and Christina Neumayer, "Laying Claim to Social Media by Activists: A Cyber-material Detournement," *Social Media + Society* 2, no. 3 (2016): 1–14.

68. Velkova and Kaun, "Algorithmic Resistance: Media Practices and the Politics of Repair."

69. Gangneux, "Tactical Agency? Young People's (Dis) Engagement with WhatsApp and Facebook Messenger."

70. Kant, *Making It Personal*.

71. Kant, *Making It Personal*, 215.

72. Stefania Milan, "When Algorithms Shape Collective Action: Social Media and the Dynamics of Cloud Protesting," *Social Media + Society* 1, no. 2 (July–December 2015): 1–10.

73. Taina Bucher, *If . . . Then: Algorithmic Power and Politics* (Oxford: Oxford University Press, 2018).

74. Claudia Wagner, Markus Strohmaier, Alexandra Olteanu, Emre Kıcıman, Noshir Contractor, and Tina Eliassi-Rad, "Measuring Algorithmically Infused Societies," *Nature* 595, no. 7866 (2021): 197–204.

75. Cotter, "Playing the Visibility Game."

76. Isabella Rega and Andrea Medrado, "The Stepping into Visibility Model: Reflecting on Consequences of Social Media Visibility—A Global South Perspective," *Information, Communication & Society* (2021): 1–20. DOI: 10.1080/1369118X.2021.1954228.

77. Guobin Yang, "Narrative Agency in Hashtag Activism: The Case of# BlackLivesMatter," *Media and Communication* 4, no. 4 (2016): 13–17.

78. Zeynep Tufekci, *Twitter and Tear Gas: The Power and Fragility of Networked Protest* (New Haven, CT: Yale University Press, 2017).

79. For an updated review of cases around the world, see Woolley, *The Reality Game*; Samuel C. Woolley and Philip N. Howard, "Computational Propaganda Worldwide: Executive Summary," Oxford University working paper no. 2017.11 (2017).

80. Samuel C. Woolley and Douglas Guilbeault, "Computational Propaganda in the United States of America: Manufacturing Consensus Online," Oxford University working paper no. 2017.5 (2017).

81. Paula Ricaurte Quijano, "Tan cerca de Twitter y tan lejos de los votantes: las estrategias de los candidatos presidenciales mexicanos durante la campaña electoral de 2012," *Versión: Estudios de Comunicación y Política* 31 (2013): 90–104.

82. Jacinta Mwende Maweu, "Still Manufacturing Consent in the Digital Era: Disinformation, 'Fake News' and Propaganda in the 2017 Elections in Kenya," in *Propaganda in the Information Age*, ed. Alan MacLeod (London: Routledge, 2019), 141–153.

83. Tanya Filer and Rolf Fredheim, "Popular with the Robots: Accusation and Automation in the Argentine Presidential Elections, 2015," *International Journal of Politics, Culture, and Society* 30, no. 3 (2017): 259–274.

84. Filer and Fredheim, "Popular with the Robots, 2015."

85. Shyam S. Sundar, Anne Oeldorf-Hirsch, and Qian Xu, "The Bandwagon Effect of Collaborative Filtering Technology," in *Proceeding of the Twenty-Sixth Annual CHI Conference Extended Abstracts on Human Factors in Computing Systems—CHI '08*, eds. Mary Czerwinski, Anne Lund, and Desney Tan (New York: ACM Press, 2008), 3453–3458.

86. Tobias R. Keller and Ulrike Klinger, "Social Bots in Election Campaigns: Theoretical, Empirical, and Methodological Implications," *Political Communication* 36, no. 1 (2019): 171–189

87. Sarah J. Jackson, Moya Bailey, and Brooke Foucault Welles, *#HashtagActivism: Networks of Race and Gender Justice* (Cambridge, MA: MIT Press, 2020).

88. Treré, *Hybrid Media Activism: Ecologies, Imaginaries, Algorithms*; Treré, "From Digital Activism to Algorithmic Resistance."

89. see also Ramón Feenstra, Simon Tormey, Andreu Casero-Ripollés, and John Keane, *Refiguring Democracy: the Spanish Political Laboratory* (New York: Routledge, 2017).

90. Paolo Gerbaudo, "From Cyber-autonomism to Cyber-populism: An Ideological History of Digital Activism," *TripleC: Communication, Capitalism & Critique* 15, no. 2 (2017): 477–489.

91. Jose Candón-Mena and Emiliano Treré. "Visionarios pragmáticos: Imaginarios, mitos y tecnopolítica en el movimiento 15M," *Revista Española de Investigaciones Sociológicas: Reis* (2022).

92. Yoav Halperin, "Reclaiming the People: Counter-Populist Algorithmic Activism on Israeli Facebook," *Television & New Media* (January 2022): 1–17

93. Halperin, "Reclaiming the People: Counter-Populist Algorithmic Activism on Israeli Facebook," 8.

94. Bucher, *If . . . Then: Algorithmic Power and Politics.*

95. Halperin, "Reclaiming the People: Counter-Populist Algorithmic Activism on Israeli Facebook," 9.

96. Gavin Haynes, "Search 'Idiot,' Get Trump: How Activists Are Manipulating Google Images," *The Guardian*, July 17, 2018, https://www.theguardian.com/us -news/2018/jul/17/trump-idiot-google-images-search.

97. See Reddit, https://www.reddit.com/r/The_Mueller/comments/8on76n/the_first _picture_that_comes_up_when_you_google/.

98. Noble, *Algorithms of Oppression: How Search Engines Reinforce Racism.*

99. Velkova and Kaun, "Algorithmic Resistance."

100. Velkova and Kaun, "Algorithmic Resistance," 523–524.

101. Rachel E. Moran, Kolina Koltai, Izzi Grasso, Joseph Schafer, and Connor Klentschy, "Content Moderation Avoidance Strategies," Virality Project, July 29, 2021, https:// www.viralityproject.org/rapid-response/content-moderation-avoidance-strategies-used -to-promote-vaccine-hesitant-content.

102. Stevie Chancellor, Jessica Pater, Trustin Clear, Eric Gilbert, and Munmun De Choudhury, "#thyghgapp: Instagram Content Moderation and Lexical Variation in Pro-Eating Disorder Communities," Georgia Institute of Technology, 2016, http://www.munmund.net/pubs/cscw16_thyghgapp.pdf; Ysabel Gerrard, "Beyond the Hashtag: Circumventing Content Moderation on Social Media," *New Media & Society* 20, no.12 (2018): 4492–4511.

103. Prashanth Bhat and Ofra Klein, "Covert Hate Speech: White Nationalists and Dog Whistle Communication on Twitter," in *Twitter, the Public Sphere, and the Chaos of Online Deliberation*, eds. Gwen Bouvier and Judith E. Rosenbaum (London: Palgrave Macmillan, 2020), 151–172.

104. Moran, Koltai, Grasso, Schafer, and Klentschy, "Content Moderation Avoidance Strategies."

105. Chancellor, Pater, Clear, Gilbert, and De Choudhury, "#thyghgapp: Instagram Content Moderation and Lexical Variation in Pro-Eating Disorder Communities."

106. Bhat and Klein, "Covert Hate Speech: White Nationalists and dog Whistle Communication on Twitter."

107. Ozlem Demirkol Tønnesen, "'Two Can Play at That Game': Communicating Dissent as a Micro-celebrity in a Restricted National Twittersphere," Selected Paper of #AoIR2020: The 21st Annual Conference of the Association of Internet Researchers Virtual Event / October 27–31, 2020.

108. Tønnesen, "'Two Can Play at That Game,'" 2–3.

109. We noticed this encryption tactic after a long period of observation of some of the most famous profiles of exponents of the Italian extreme right online, which lasted between spring 2021 and spring 2022. We also wrote a private message to some of them, on Twitter, to ask them why they encrypt some of their tweets, and this was the answer that one of them gave: "I do this simply to prevent being banned by Twitter and to defend myself from the maniacal trolls who do Twitter searches based on 'banned' keywords in order to find 'offensive' tweets and report them to Twitter. I've only been banned once because I wrote 'Hitler would have been proud of it' commenting on a repressive measure but there are groups of lunatics that every day search for 'Hitler' and report who writes it." Another user also told us that they like encrypting messages on Twitter because they feel to belong to a kind of "secret society." To learn more about so-called algospeak and its dangerous implications, see Alexandra S. Levine, "From Camping to Cheese Pizza, 'AlgoSpeak' Is Taking over Social Media," *Forbes*, September 19, 2022. https://www.forbes.com/sites/alexandralevine/2022/09/16/algospeak-social-media-survey/?sh=2b95197455e1.

110. The first comment below the tweet accuses Michele Boldrin of being a communist: "Note also that during a conflict between Russia and Ukraine, he brings up small Italian companies. What a communist!" The second comment says: "Let's just call him a dickhead."

111. Finn Brunton and Helen Nissenbaum, "Vernacular Resistance to Data Collection and Analysis: A Political Theory of Obfuscation," *First Monday* (2011), https://firstmonday.org/ojs/index.php/fm/article/download/3493/2955.

112. Brunton and Nissenbaum, "Vernacular Resistance to Data Collection and Analysis."

113. João C. Magalhães, "Algorithmic Resistance as Political Disengagement," *Media International Australia* 183, 1 (2022): 77–89.

114. "Algorithmic Resistance as Political Disengagement," 79.

115. Nikita Jain, Pooja Agarwal, and Juhi Pruthi, "Hashjacker—Detection and Analysis of Hashtag Hijacking on Twitter," *International Journal of Computer Applications* 114, no. 19 (March 2015): 17–20.

116. Tommaso Gravante, "Forced Disappearance as a Collective Cultural Trauma in the Ayotzinapa Movement," *Latin American Perspectives* 47, no. 6 (2020): 87–102.

117. Erin Gallagher, "Mexico: Articles about Bots and Trolls," *Medium*, January 1, 2017, https://erin-gallagher.medium.com/news-articles-about-bots-trolls-in -mexican-networks-7b1e551ef4a6.

118. Klint Finley, "Pro-Government Twitter Bots Try to Hush Mexican Activists," Wired.com, August 23, 2015, https://www.wired.com/2015/08/pro-government -twitter-bots-try-hush-mexican-activists/.

119. See Alberto Escorcia's LoQueSigue (WhatFollows) website (https://loquesigue.tv/).

120. Sarah J. Jackson and Brooke Foucault Welles, "Hijacking# myNYPD: Social Media Dissent and Networked Counterpublics," *Journal of Communication* 65, no. 6 (2015): 932–952.

121. Aja Romano, "How K-pop Fans Are Weaponizing the Internet for Black Lives Matter," Vox.com, June 22, 2020, https://www.vox.com/2020/6/8/21279262/k-pop -fans-black-lives-matter-fancams-youtubers-protest-support.

122. Crystal Abidin and Thomas Baudinette, "The Civic Hijinks of K-Pop's Super Fans," *Data & Society*, July 1, 2020, https://points.datasociety.net/the-civic-hijinks-of -k-pops-super-fans-ae2e66e28c6.

123. Abby Ohlheiser, "How K-pop Fans Became Celebrated Online Vigilantes," *MIT Technology Review*, June 5, 2020, https://www.technologyreview.com/2020/06/05 /1002781/kpop-fans-and-black-lives-matter/

124. Dexter Thomas, "Is This Beverly Hills Cop Playing Sublime's 'Santeria' to Avoid Being Live-Streamed?" *Vice*, February 9, 2021, https://www.vice.com/en/article/bvxb94 /is-this-beverly-hills-cop-playing-sublimes-santeria-to-avoid-being-livestreamed.

125. Michael Etter and Oana Brindusa Albu, "Activists in the Dark: Social Media Algorithms and Collective Action in Two Social Movement Organizations," *Organization* 28, no. 1 (2021): 68–91.

126. Dean, *Democracy and Other Neoliberal Fantasies*, 48.

CHAPTER 6

1. William Gibson, "Rocket Radio," *Rolling Stone* 15 (1989): 84.

2. Nicola Davis, "'Yeah, We're Spooked': AI Starting to Have Big Real-World Impact, Says Expert," *The Guardian*, October 29, 2021, https://www.theguardian

.com/technology/2021/oct/29/yeah-were-spooked-ai-starting-to-have-big-real-world
-impact-says-expert.

3. Bernard Stiegler, *Automatic Society, Volume 1: The Future of Work* (London: Polity
Press, 2016).

4. Zuboff, *The Age of Surveillance Capitalism.*

5. Max Horkheimer and Theodore W. Adorno, *Dialectic of the Enlightenment* (Stanford, CA: Stanford University Press, 2002), 19; quoted in Mark Andrejevic, *Automated Media* (London: Routledge, 2019), 162.

6. Andrejevic, *Automated Media.*

7. Andrejevic, *Automated Media*, 6.

8. Dean, *Democracy and Other Neoliberal Fantasies.*

9. Dean, *Democracy and Other Neoliberal Fantasies*, 23.

10. Delfanti, *The Warehouse. Workers and Robots at Amazon.*

11. Antonio Casilli, *En attendant les robots: Enquête sur le travail du clic* (Paris: Seuil, 2019).

12. Milagros Miceli and Julian Posada, "The Data-Production Dispositif," *arXiv preprint arXiv:2205.11963* (2022), 4.

13. Phil Jones, *Work without the Worker: Labour in the Age of Platform Capitalism* (London: Verso Books, 2021).

14. Phil Jones, "Refugees Help Power Machine Learning Advances at Microsoft, Facebook, and Amazon," *Rest of World*, September 21, 2021. https://restofworld.org/2021/refugees-machine-learning-big-tech/.

15. Jones, "Refugees Help Power Machine Learning Advances at Microsoft, Facebook, and Amazon."

16. Billy Perrigo, "Exclusive: OpenAI Used Kenyan Workers on Less than $2 per Hour to Make ChatGPT Less Toxic," *Time*, January 18, 2023, https://time.com/6247678/openai-chatgpt-kenya-workers/.

17. Kate Crawford, *The Atlas of AI* (New Haven, CT: Yale University Press, 2021).

18. Nick Dyer-Witheford, Atle Mikkola Kjøsen, and James Steinhoff, *Inhuman Power. Artificial Intelligence and the Future of Capitalism* (London: Pluto Press, 2019).

19. Andreas Hepp, *Deep Mediatization* (London: Routledge, 2019).

20. Mark Deuze, *Media Life* (Cambridge, UK: Polity Press, 2012).

21. See van Dijck et al., *Platform Society*. See also José van Dijck, "Seeing the Forest for the Trees: Visualizing Platformization and Its Governance," *New Media & Society* 23, no. 9 (2021): 2801–2819.

22. Plantin et al., "Infrastructure Studies Meet Platform Studies in the Age of Google and Facebook."

23. Vasudevan and Chan, "Gamification and Work Games: Examining Consent and Resistance among Uber Drivers." See also Ngai Keung Chan, *Managing Algorithmic*

Metrics and Customers: A Multi-Case Study of Labor Control and Resistance in the Gig Economy, PhD dissertation, Cornell University, Ithaca, NY, 2021.

24. Steve Hendrix, "Traffic-Weary Homeowners and Waze Are at War, Again. Guess Who's Winning?" *Washington Post*, June 5, 2016, https://www.washingtonpost.com /local/traffic-weary-homeowners-and-waze-are-at-war-again-guess-whos-winning /2016/06/05/c466df46-299d-11e6-b989-4e5479715b54_story.html.

25. Brian Barrett, "An Artist Used 99 Phones to Fake a Google Maps Traffic Jam," *Wired*, February 3, 2020, https://www.wired.com/story/99-phones-fake-google-maps -traffic-jam/.

26. Oobah Butler, "I Made My Shed the Top Rated Restaurant On TripAdvisor," *Vice*, December 6, 2017, https://www.vice.com/en/article/434gqw/i-made-my-shed -the-top-rated-restaurant-on-tripadvisor.

27. Scott, *Weapons of the Weak*, 299; "Many of the forms of resistance I have been examining may be individual actions, but this is not to say that they are uncoordinated."

28. Pyotr Alekseevič Kropotkin, *Mutual Aid: A Factor of Evolution* (New York, McClure Phillips & Co., 1902).

29. Stark and Pais, "Algorithmic Management in the Platform Economy"; Francesco Bonifacio, "Cycling as a Food-Delivery Rider. Or the Difficult Negotiation among Speed, Safety and Accuracy," *Eracle: Journal of Sport and Social Sciences* 5, no. 1 (2022): 148–164.

30. Kenan Malik, "In an Age Too Given to Moral Certainty, Let's Remember The Wire's Omar as a Study in Complexity," *The Guardian*, September 12, 2021, https:// www.theguardian.com/commentisfree/2021/sep/12/in-a-black-and-white-age-the -wire-omar-little-illuminated-our-knotty-moral-lives?.

31. A Spotify software developer whom we interviewed in 2018 believes that the platform user is in control of her listening experience because "everything is based on your listening habits in Spotify." From this series of interviews, we published the following article: Bonini and Gandini, "'First Week Is Editorial, Second Week Is Algorithmic': Platform Gatekeepers and the Platformization of Music Curation."

32. Tarleton Gillespie, "Can an Algorithm Be Wrong?" *Limn* 1, no. 2 (2012). https:// limn.it/articles/can-an-algorithm-be-wrong/. See also Dan Milmo, "Twitter Admits Bias in Algorithm for Rightwing Politicians and News Outlets," *The Guardian*, October 22, 2021, https://www.theguardian.com/technology/2021/oct/22/twitter-admits -bias-in-algorithm-for-rightwing-politicians-and-news-outlets.

33. Nick Seaver, "Algorithms as Culture: Some Tactics for the Ethnography of Algorithmic Systems," *Big Data & Society* 4, no. 2 (2017): 2053951717738104.

34. Judy Wajcman, "How Silicon Valley Sets Time," *New Media & Society* 21, no. 6 (2019): 1276.

35. Winner, "Do Artifacts Have Politics?"

36. Adrienne Shaw, "Encoding and Decoding Affordances: Stuart Hall and Interactive Media Technologies," *Media, Culture & Society* 39, no. 4 (2017): 592–602.

37. Lomborg and Kapsch, "Decoding Algorithms."

38. Hollander and Einwohner, "Conceptualizing Resistance."

39. Scott, *Weapons of the Weak*, 301.

40. Stuart Hall and Tony Jefferson, eds., *Resistance through Rituals* (London: Harper Collins, 1976).

41. James Procter, *Stuart Hall* (London: Routledge, 2004), 90 (emphasis in original).

42. Stuart Hall, John Clarke, Tony Jefferson, and Brian Roberts, "Subcultures, Cultures and Class: A Theoretical Overview," in *Resistance through Rituals*, eds. Stuart Hall and Tony Jefferson (Routledge, 2006), 47.

43. Edward P. Thompson, *The Making of the English Working Class* (London: Vintage, 1966).

44. Here, we refer to the metaphor of bits and atoms created by Nicholas Negroponte. With the word "atoms," he intended the physical entity, while "bits" stand for the emerging cyberspace (in 1995). See Nicholas Negroponte, *Being Digital* (New York: Alfred A. Knopf, 1995).

45. Thompson, *The Making of the English Working Class*, 424.

46. Nieborg, Poell, and Duffy, "Analyzing Platform Power in the Cultural Industries."

47. Thompson, *The Making of the English Working Class*, 424 (emphasis in original).

48. Rida Qadri and Noopur Raval, "Mutual aid Stations," *Logic* 13 (May 2021). https://logicmag.io/distribution/mutual-aid-stations/.

49. Qadri and Raval, "Mutual Aid Stations."

50. Scott, *Weapons of the Weak*, 297.

51. Eric J. Hobsbawm "The Machine Breakers," *Past & Present* 1 (February 1952), 58; quoted in Mueller, *Breaking Things at Work*, 16.

52. Hobsbawm, "The Machine Breakers," 58.

53. Steve Wright, *Storming Heaven: Class Composition and Struggle in Italian Autonomist Marxism* (London: Pluto Press, 2002).

54. See also Giuseppina Mecchia and Max Henninger, "Introduction: Italian Post-Workerist Thought," *SubStance* 36, no. 1 (2007): 3–7.

55. Toni Negri, "Pour une définition ontologique de la multitude," *Multitudes* 2 (2005): 36–48.

56. Zuboff, "The Age of Surveillance Capitalism."

57. TNM Staff, "Auto Unions in Bengaluru to Launch Their Own Ridesharing App," *The News Minute*, October 10, 2022. https://www.thenewsminute.com/article/auto-unions-bengaluru-launch-their-own-ridesharing-app-168726.

58. Ivan Illich, *Tools for Conviviality* (London: Calder and Boyars, 1973).

59. See Adam Arvidsson, "Capitalism and the Commons," *Theory, Culture & Society* 37, no. 2 (2020): 3–30.

60. Adam Arvidsson, *Changemakers: The Industrious Future of the Digital Economy* (London: Polity, 2019).

61. See, for example, the work of Algorithm Watch, a Berlin-based, nonprofit research and advocacy organization that "is commited to watch, unpack and analyze automated decision-making (ADM) systems and their impact on society." See: https://algorithmwatch.org/en/. Other organizations and research centers include the Data Justice Lab at Cardiff University, which examines the intricate relationship between datafication and social justice, highlighting the politics and impacts of data-driven processes and big data (https://datajusticelab.org/) and the Algorithmic Justice League, the Mozilla Foundation, the Ada Lovelace Institute, Data for Black Lives, the DAIR Institute, the AI on the Ground Initiative, the BigDataSur Initiative, Signa_Lab, and other organizations.

62. Scott, *Weapons of the Weak*, 293.

63. Scott, *Weapons of the Weak*, 350.

64. Hall et al., "Subcultures, Cultures and Class," 44 (emphasis in original). The original sentences read: "Negotiation, resistance, struggle: The relations between a subordinate and a dominant culture, wherever they fall within this spectrum, are always intensively active, always oppositional, in a structural sense. Their outcome is not given but *made*."

APPENDIX

1. Dorismilda Flores-Marquez and Rodrigo González Reyes, eds., "La imaginación metodológica: coordenadas, rutas y apuestas para el estudio de la cultura digital," *Ciudad de México: Tintable* (2021).

2. On the challenges and opportunities of doing research during the pandemic, see Michael D. Fetters and José F. Molina-Azorin, "Special Issue on COVID-19 and Novel Mixed Methods Methodological Approaches during Catastrophic Social Changes," *Journal of Mixed Methods Research* 15, no. 3 (2021): 295–303; Stephanie Tremblay, Sonia Castiglione, Li-Anne Audet, Michèle Desmarais, Minnie Horace, and Sandra Peláez, "Conducting Qualitative Research to Respond to COVID-19 Challenges: Reflections for the Present and Beyond," *International Journal of Qualitative Methods* 20 (2021), https://doi.org/10.1177/16094069211009679; Ash Watson and Deborah Lupton, "Remote Fieldwork in Homes during the COVID-19 Pandemic: Video-Call Ethnography and Map Drawing Methods," *International Journal of Qualitative Methods* 21 (2022), https://doi.org/10.1177/16094069221078376.

3. Bogusia Temple, "Watch Your Tongue: Issues in Translation and Cross-cultural Research," *Sociology*, 31, no. 3 (1997): 607–618.

4. Ana Cristina Suzina, "English as *Lingua Franca*: Or the Sterilisation of Scientific Work," *Media, Culture & Society* 43, no. 1 (2021): 171–179.

5. Sara Pink, Heather Horst, John Postill, Larissa Hjorth, Tania Lewis, and Jo Tacchi, *Digital Ethnography: Principles and Practice* (London: SAGE, 2015).

6. Sérgio Barbosa and Stefania Milan, "Do Not Harm in Private Chat Apps: Ethical Issues for Research on and with WhatsApp," *Westminster Papers in Communication and Culture* 14, no. 1 (2019): 49–65.

7. Woodcock, *The Fight against Platform Capitalism.*

8. Barbosa and Milan, "Do Not Harm in Private Chat Apps," 53.

9. A first version of this project was presented at the ECREA conference in Helsinki 2019; at the preconference of AoIR 2020 in Dublin, entitled "Resisting Algorithms"; in December 2020 in a webinar entitled "How to Strike" and organized by the Italian Association of Art Workers; and in various keynotes, seminars, and events during 2020 and 2021 in various Latin American countries (including Mexico, Colombia, and Ecuador).

10. Zizheng Yu, Emiliano Treré, and Tiziano Bonini, "The Emergence of Algorithmic Solidarity: Unveiling Mutual Aid Practices and Resistance among Chinese Delivery Workers," *Media International Australia* (2022): 1329878X221074793.

11. Tiziano Bonini, Emiliano Treré, Zizheng Yu, Swati Singh, Daniele Cargnelutti, and Francisco Javier López-Ferrández, "Cooperative Affordances: How Instant Messaging Apps Afford Learning, Resistance and Solidarity among Food Delivery Workers," *Convergence,* 0, no. 0 (2023).

12. Tiziano Bonini, Emiliano Treré, and Francesca Murtula, "Resistenza e solidarietà algoritmica nelle piattaforme digitali: un'indagine etnografica dei gruppi di engagement su Instagram," *Studi Culturali,* 19, no. 2 (2022): 177–206.

13. Emiliano Treré and Tiziano Bonini, "Amplification, Evasion, Hijacking: Algorithms as Repertoire for Social Movements and the Struggle for Visibility," *Social Movement Studies* (2022). DOI: 10.1080/14742837.2022.2143345.

14. See, for example, Deborah Lupton, ed., "Doing Fieldwork in a Pandemic," crowd-sourced document, 2020, https://docs.google.com/document/d/1clGjGABB2h2qbdu TgfqribHmog9B6P0NvMgVuiHZCl8/edit?ts=5e88ae0a; Marnie Howlett, "Looking at the 'Field' through a Zoom Lens: Methodological Reflections on Conducting Online Research during a Global Pandemic," *Qualitative Research* (2021): 1468794120985691; Melanie Nind, Andy Coverdale, and Robert Meckin, "Changing Social Research Practices in the Context of Covid-19: Rapid Evidence Review," National Centre for Research Methods, Southampton, UK, 2021, https://eprints.ncrm.ac.uk/id/eprint /4457/; Yenn Lee, "The NCRM Wayfinder Guide to Conducting Ethnographic Research in the COVID-19 Era," National Centre for Research Methods, Southampton, UK, 2021, https://eprints.ncrm.ac.uk/id/eprint/4410/; Angèle Christin, "Algorithmic Ethnography during and after COVID-19," *Communication and the Public* 5, no. 3–4 (2020): 108–111; Magdalena H. Góralska, "Anthropology from Home: Advice on Digital Ethnography for the Pandemic Times," *Anthropology in Action* 27, no. 1 (2020): 46–52.

15. See Pink et al., *Digital Ethnography: Principles and Practice*; Alessandro Caliandro, "Digital Methods for Ethnography: Analytical Concepts for Ethnographers Exploring Social Media Environments," *Journal of Contemporary Ethnography* 47, no. 5 (2018): 551–578; Crystal Abidin and Gabriele De Seta, "Private Messages from the Field: Confessions on Digital Ethnography and Its Discomforts," *Journal of Digital Social Research* 2, no. 1 (2020): 1–19.

16. Angèle Christin, "Algorithmic Ethnography during and after COVID-19," 108.

17. Angèle Christin, "Algorithmic Ethnography during and after COVID-19," 109.

18. Angèle Christin, "The Ethnographer and the Algorithm: Beyond the Black Box," *Theory and Society* 49, no. 5 (2020): 897–918.

19. Angèle Christin, "Algorithmic Ethnography during and after COVID-19," 108.

20. George E. Marcus, "Ethnography in/of the World System: The Emergence of Multi-sited Ethnography," *Annual Review of Anthropology* 24, no. 1 (1995): 95–117.

21. Ulf Hannerz, "Being There . . . and There . . . and There! Reflections on Multi-site Ethnograph," *Ethnography* 4, no. 2 (2003): 201–216.

22. Gabriele de Seta, "Three Lies of Digital Ethnography," *Journal of Digital Social Research* 2, no. 1 (2020): 77–97.

23. de Seta, "Three Lies of Digital Ethnography," 81.

24. Nicholas W. Jankowski and Martine van Selm, "Epilogue: Methodological Concerns and Innovations in Internet Research," in *Virtual Methods: Issues in Social Research on the Internet*, ed. Christine Hine (Oxford, UK: Berg, 2005), 199–207.

25. Clare Madge, "Developing a Geographers' Agenda for Online Research Ethics," *Progress in Human Geography* 31 (2007): 654–674.

26. Edgar Gómez-Cruz and Ignacio Siles, "Visual Communication in Practice: A Texto-material Approach to WhatsApp in Mexico City," *International Journal of Communication* 15, no. 21 (2021), 4546–4566. https://ijoc.org/index.php/ijoc/article/view/17503/3579.

27. Jan Švelch, "Redefining Screenshots: Toward Critical Literacy of Screen Capture Practices," *Convergence* 27, no. 2 (2021): 554–569.

28. Of course, before sharing or archiving a screenshot, we obscured any information that could reveal the identities of the people involved in the interaction.

29. Lupton, "Doing Fieldwork in a Pandemic."

30. Edgar Gómez-Cruz and Ramaswami Harindranath, "WhatsApp as 'Technology of Life': Reframing Research Agendas," *First Monday* 25, no. 12 (2020). https://doi.org/10.5210/fm.v25i12.10405.

31. Kathy Charmaz, *Constructing Grounded Theory: A Practical Guide through Qualitative Analysis* (London: SAGE, 2006).

32. Tiziano Bonini and Francesca Murtula, "'Ancora non ci ho capito niente di come funziona l'algoritmo': La consapevolezza algoritmica degli host di Airbnb," *Sociologia Italiana* 19–20 (2022): 147–161. Research for this article was financed by the Italian Research funding programme PRIN ("Projects of Significant National Interest"): PRIN 2017. Prot. 2017EWXN2F. "The Short-Term City: Digital Platforms and Spatial (In)justice" research project [STCity].

INDEX

Page numbers in italic indicate figures.

Abidin, Crystal, 120, 153, 208n22
Activism, algorithmic
    agency and resistance in, 161–165
    concept of, 10, 137–142
Ada Lovelace Institute, 216n60, 224n61
Adler-Bell, Sam, 75
Adorno, Theodore W., 156
Agency. *See* Algorithmic agency
Agnosticism of algorithmic politics, 10,
    139, 154
AI on the Ground Initiative, 216n60,
    224n61
Airbnb
    algorithmic agency and resistance to,
    50–52, *51f*, 55, *171f*
    moral economy of, 36
    platform labor exploited by, 5, 62
    research methodology and, 188
Airoldi, Massimo, 19
Algoactivism, 216n50
Algorithm-enabled resistance, 24, 139,
    169

Algorithmic activism, 137–142
    agency and resistance of, 161–165
    agnosticism of, 10, 139, 154
    data-enabled versus data-oriented,
    139–140
    moral economy of, 10, 143–147
Algorithmic agency, 16–18
    algorithmic alliances, 56–57, 76–92
    common characteristics of, 161–165
    concept of, 2–4, 7, 18–20
    in cultural industries. *See* cultural
    industries
    of food delivery couriers. *See* food
    delivery couriers
    free versus platform labor, 4–6
    manifestations of, *31f*, 47–56,
    170–171, *171f*, 198n65
    moral dimension of. *See* moral
    economies
    in politics. *See* politics, algorithmic
    power relations of, 4–6, 57–58
    productive, 24–25

Algorithmic agency (cont.)
  relevance of, 161–165
  strategic versus tactical, 43–47
  symbiotic nature of, 19–20
Algorithmic alliances, 56–57, 76–92
Algorithmic amplification, 10, 143–147, 154
Algorithmic bias/discrimination, 3–4, 15–16, 54, 140, 189n9
Algorithmic ethnography, 185–186
Algorithmic evasion, 10, 147–150, *149f*, 154, 219n109
Algorithmic gossip, 198n73
Algorithmic hijacking, 150–154
Algorithmic imaginaries, 47, 95, 198n73
Algorithmic Justice League, 216n60, 224n61
Algorithmic politics. *See* Politics, algorithmic
Algorithmic populism, 133, 135
Algorithmic resistance, 56. *See also* Algorithmic agency
  automation of society and, 155–161
  banality of, 26–27
  concept of, 7, 20–23, 144
  in cultural industries. *See* cultural industries
  everyday forms of, 25–27, 169–171
  of food delivery couriers. *See* food delivery couriers
  intentionality of, 22
  microresistance, 25
  in politics. *See* politics, algorithmic
  productivity of, 24–25
  relevance of, 169–171, *171f*
  resistance *through* versus resistance *to* algorithms, 23–25, 169–170
Algorithmic Resistance Project (AlgoRes), 180–183
Algorithmic solidarity. *See* Solidarity
Algorithm Watch, 216n60, 224n61
Algospeak, 219n109

Alibaba, 13, 114
Alliances, algorithmic, 56–57, 76–92
All-India Gig Workers' Union (AIGWU), 97
Amazon, 13, 62. *See also* Platform power
  algorithmic workforce management, 75–76, 157
  Mechanical Turk, 62
American Influencer Council, 130
Amoore, Louise, 34
Amplification, algorithmic, 10, 143–147, 154
Andrejevic, Mark, 156
Andres, Lauren, 47
Ang, Ien, 6
Apple, 13
Architects of disinformation, 137
Artificial intelligence (AI), 10, 19–20, 140, 155–161
Arvidsson, Adam, 177
Assodelivery, 202n23
Astroturfing, 53, 135, 214n25
Atoms, metaphor of, 171, 223n44
Automated booking of working shifts, 79, *80f*
Automation, 10
  algorithmic agency, relevance of, 161–165
  algorithmic resistance, relevance of, 155–161, 169–171, *171f*
  consequences of, 155–161
  moral economy, relevance of, 165–169
  platform working class, making of, 171–178
  power balances in, 160–161
  uncertain outcome of, 177–178
Autonomous University of Querétaro, 185
Autorickshaw Drivers Union (ARDU), 177
Ayotzinapa teachers' college, 151

Baaz, Mikael, 23
Baidu, 13
Bandwagon effect, 143
Barbosa, Sergio, 182
BAT (Baidu, Alibaba, and Tencent), 13
Baudinette, Thomas, 153
Becker, Howard, 12
Behavioral surplus accumulation, 176
Benefit societies, 172–174
Benjamin, Ruha, 140
Beraldo, Davide, 23, 133, 139
Berg, Anne Jorunn, 34
Bhat, Prashanth, 148
Bias/discrimination, algorithmic, 3–4,
    15–16, 54, 140, 189n9
BigDataSur Initiative, 224n61
Bishop, Sophie, 198n73
BitTorrent, 150
#BlackLivesMatter hashtag, 142, 144
Blomkamp, Neill, 160
#BlueLivesMatter hashtag, 153
Bolt, 98–99
Bonelli, Susanna, 188
Bonifacio, Francesco, 82
Bots, 41
    algorithmic hijacking by, 150–153
    in cultural industries, 116, 120, 122,
        124
    food delivery courier use of, 77, 79,
        101
    political, 135, 143, 151–0152
    strategic algorithmic agency and, 46,
        52–53
Botxo couriers, 104
Bread revolt, moral economy of, 31–33
Brexit, 134
Brunton, Finn, 150
BTS fan communities, tactical algorith-
    mic agency of, 107–108, 115–117
Buchanan, Ian, 44, 58
Bucher, Taina, 18, 113, 198n73
Burai, Johanna, 147
Burrell, Jenna, 34

Cabañes, Jason Vincent, 37, 100, 129,
    137
Cabiify, 181
Cambridge Analytica, 134
Canadian school, 15
Capitalism
    communicative, 36, 157
    digital, 15–16, 156, 159, 176
    industrial, 73–74, 172
    platform, 10, 36, 73–74, 157, 159,
        172–173
    surveillance, 14, 17, 58, 81, 156
Cardiff University, 185, 224n61
Cargnelutti, Daniele, 181, 182
Casa del Courier (Couriers' Home), 98
Casa del Rider, 173
Casilli, Antonio, 157
Center for Human-Compatible Artificial
    Intelligence, 155
Chakravartty, Paula, 136
Chan, Ngai Keung, 206n81
Chaos of platform society, 3, 17
Charlesworth, Andrew, 33, 104
Chat groups, 91–92
    as hidden transcripts of resistance,
        96–97
    as learning environments, 93–96,
        163–164
    as mutual aid and solidarity-building
        spaces, 97–99, 163–164, 173
Chen, Julie Yujie, 67
Christin, Angèle, 185–186, 216n50
Civic hacktivism, 24–25
Cloud protesting, 25
Collective action, 10. See also Chat
    groups; Unions
    in algorithmic politics, 137–142
    among gig workers, 85–92
    in cultural industries, 117–130, 125f,
        126f
    definition of, 76
    moral economies and, 37
    relevance of, 161–165

Collective intelligence, 94
Colonialism, data, 3, 7, 15
Communicative capitalism, 36, 157
Competing moral economies
  in algorithmic politics, 143–147
  among gig workers, 100–104, 206n81
  in cultural industries, 36, 41, 122–127,
    126f, 127f, 168–169
  platform paternalism and, 38–40
Computational propaganda, 10, 41, 46,
  53, 133, 135, *171f*
Computation power, food delivery
  platforms, 63–67
Confederazione Generale Italiana del
  Lavoro (CGIL), 98
Consensus, manufacturing of, 143
Content creators. *See* Cultural industries
Content ID, 153
Contention, repertoires of, 10, 141–142
Contentious/tactical algorithmic poli-
  tics, 10, 137–142
Content moderation avoidance strate-
  gies, 147–148, 219n109
Contesting of platform affordances,
  90–91
Coop Cycle, 104–105, 207n87
Cooperative ethics, 161–165. *See also*
  Collective action
Cooperativism, platform, 104–106
Coordinated order refusal, 86–87
Copyleft, 207n87
Copyright protection algorithm,
  131
Costanza-Chock, Sasha, 140
Couldry, Nick, 3, 19
Courier-led *Shuadan,* 83–84
Couriers, food delivery. *See* Food
  delivery couriers
Courpasson, David, 21, 22
COVID-19 pandemic, 1–2, 114, 134,
  179–180
Crawford, Kate, 158
The Creator Union (TCU), 130

Cultural artifacts
  algorithms as, 167–169
  screenshots as, 187–188
Cultural industries
  competing moral economies of,
    122–127
  digital labor exploited by, 4–6
  engagement groups in, 117–130
  folk theories of social feeds in, 95, 120
  gaming practices in, 8–9, 107–108,
    114–117
  platformization of, 9, 14, 108–114,
    123
  platform working class, making of,
    171–178
  relevance of, 161–165
  research methodology and, 184
  Sajaegi, 116–117
  solidarity and collectivist values in,
    173–177
  sumseuming, 115–117
  unions in, 129–130
  visibility in, 9, 111–114, 142–143, 164,
    208n22
Cultural optimization, 48–50, *171f*
Cyberautonomism, 146
Cyber-material détournement, 141
Cyberpopulism, 146

DAIR Institute, 216n60, 224n61
Dallas Police Department, 152–153
Dark participation, 134
Data activism, 17, 139
Data agency, 17, 139
Data as repertoire, 23, 139
Data as stakes, 23, 139
Data colonialism, 3, 7, 15, 17
Data-enabled activism, 23, 139–140
Datafication. *See also* Gamification
  data politics and, 61–71, 78, 133,
    138–139
  definition of, 3, 5
  gig economy and, 18

Data for Black Lives, 216n60, 224n61
Data Justice Lab, Cardiff University, 216n60, 224n61
Data-oriented activism, 23, 139–140
Data work, 157–158
Davis, Jennie, 90–91
Davis, Thomas, 180, 185
Dean, Jodi, 154, 157
de Balzac, Honoré, 111
de Certeau, Michel, 7, 44–45, 47, 54, 58, 82, 133, 141, 169
#DeclineNow hashtag, 86–87
Decoding, 40, 42
Decolonial turn, 16, 140
de Jong, Merit, 16
Delhi University, 181
Deliveroo, 1–2, 181, 204n43. See also Online food delivery platforms
  algorithmic agency and resistance to, 73, 79, 86–88, 103
  algorithmic alliances with, 56
  computation power of, 63–65, 168
  history of, 63
  moral economy of, 36, 168–169
  ranking systems and gamification, 70–71, 95
Delivery drivers. See Amazon; Food delivery couriers
della Porta, Donatella, 73
del Mazo, Alfredo, 52
Deplatforming, 35, 124
de Seta, Gabriele, 186
Diaries, as "technologies of the self," 77–78, 78f
Didi Food, 181
Digital capitalism, 15–16, 156, 159, 176
Digital ethnography. See Ethnographic research
Digital gig economy, 62
Digital network repertoires, 141–142
Digital values, 35. See also Moral economies

Digital work platforms. See also Cultural industries; Online food delivery platforms; Politics, algorithmic
  chaos of, 3, 17
  digital labor exploited by, 4–6
  platform power of, 3–4
  workers impacted by, 1–2
DingTalk, 114
Direct message (DM) groups. See Engagement groups
Discrimination, algorithmic, 3–4, 15–16, 54, 140, 189n9
Disengagement, algorithmic evasion and, 150
Disinformation, 134–137
Dispatchers, food delivery platforms, 66f
Dispatchers, online food delivery platforms, 65–67
Dolata, Ulrich, 138
Dominant/hegemonic decoding, 40, 42
DoorDash, 63, 86–87. See also Online food delivery platforms
Douyin, 99
Duffy, Brooke Erin, 108, 112–113, 123
Dune (film), 159–160
Dyer-Witheford, Nick, 158

EasyTaxi, 181
ECREA conference, 225n9
Edelman, Marc, 33
Einwohner, Rachel, 169
Ele.me, 63, 68, 69f, 181. See also Online food delivery platforms
Elmer, Greg, 14
Elysium (film), 160
Engagement groups, 9, 26, 108
  competing moral economies of, 122–127, 125f, 126f, 127f, 211n63
  emergence of, 117–119, 118f
  entrepreneurial solidarity in, 127–130
  gift economy in, 128
  how they work, 119–121

Engagement groups (cont.)
  tactics to avoid detection, 121–122
  types of, 120
Entrepreneurial solidarity
  among gig workers, 99–100
  in cultural industries, 127–130
  definition of, 37
Eraman, 104
Escorcia, Alberto, 151–152
Eslami, Motahhare, 120
Ethnographic research
  digital/algorithmic, 185–186
  food delivery platforms, 182–183
  gaming culture, 184
  hybrid, 185–186
  multisited, 186
Ettlinger, Nancy, 17, 24–25
Eubanks, Virginia, 4, 140
Evasion, algorithmic, 10, 147–150, *149f*,
  154

Facebook groups, 13, 91–92
  algorithmic activism on, 146
  algorithmic alliances with, 56
  algorithmic evasion on, 147–150
  algorithmic governance of, 109–110
  engagement groups, 117–130, *125f*,
    126f, 127f
  as hidden transcript of resistance,
    96–97
  as learning environments, 93–96,
    163–164
  moral economies of, 34
  as mutual aid and solidarity-building
    spaces, 97–99, 163–164, 173
  research methodology and, 181–184,
    186
  strategic algorithmic agency on, 48, 50
  visibility on, 164
Facebook Messenger, 45, 141, 143–144
Facial recognition software, 81
Fake news, 121, 134. *See also*
  Disinformation

Far-right movements, content mod-
  eration avoidance strategies of,
  146–150, *149f*, 219n109
Ferrari, Fabian, 75
15M movement, 144–145, 150
Fight Club (film), 122
First Draft News, 134
Fisher, Mark, 10–11
Flash EX, 181
Flocker, 152
Floyd, George, 152
Folk devils, 136–137
Folk theories of social feeds, 95,
  120
Food delivery couriers. *See also* Online
  food delivery platforms
  algorithmic governance of, 63–67,
    164–165
  collective tactics of, 76, 85–92
  competing moral economies of,
    100–104
  and emerging alternatives to com-
    mercial platforms, 104–106
  impact of platformization on, 1–2,
    59–61, *60f*
  individual tactics of, 76, 77–85
  motivations of, 76–77
  platform working class, making of,
    171–178
  private online chat groups used by,
    91–100, 173–175
  ranking systems and gamification and,
    67–71, *69f, 72f*
  relevance of, 161–165
  research methodology and, 11
  spatial fragmentation and, 72–73
Foodora, 64, 67. *See also* Online food
  delivery platforms
Food riots, moral economy of, 31–33
Foucault, Michel, 17, 21
Fourcade, Marion, 34
Fragmentation, food delivery couriers,
  72–73

Frank (Deliveroo), 63–65, 95, 168
Free labor, 4–6
Fuchs, Christian, 6

GAFAM (Google, Amazon, Facebook, Apple, and Microsoft), 13
Galis, Vasilis, 141
Gallagher, Erin, 151
Gamification
  online food delivery platforms, 67–71, *69f, 72f,* 80, 82, 93, 182, 206n81
  Uber, 162
Gandini, Alessandro, 5–6, 62
Gangneux, Justine, 45, 141
Gephi, 152
Gerbaudo, Paolo, 146
Gibson, William, 155
Giddens, Anthony, 7, 19–20
Gift economy, 128
Gig economy. *See also* Food delivery couriers; Online food delivery platforms
  definition of, 61–63
  digital labor exploited by, 4–6
  digital versus physical, 62
  impact of algorithms on, 1–2
  platform working class, making of, 171–178
  protests, strikes, and riots by, 71–73, 88–90
  solidarity and collectivist values in, 173–177
Gig platform couriers (GPCs), 201n20. *See also* Food delivery couriers
Gillespie, Tarleton, 14
#GirlsLikeUs hashtag, 144
Glatt, Zoe, 117, 124
Global North, 9, 12, 63
Global South, 9
  algorithmic politics of, 136–137
  digital propaganda and manipulation in, 136–137

entrepreneurial activities in, 177
gig economy in, 63, 181
mutual aid and solidary-building spaces in, 174
platform working class, making of, 171–178
WhatsApp's role in, 187–188
work automation in, 157–158
Glovo, 63–67, *66f,* 69–70, 81, 181. *See also* Online food delivery platforms
GMeet, 188
Gojek, 91
Gómez-Cruz, Edgar, 187
Google, algorithmic activism on, 13, 146–147
Governance, algorithmic, 63–67, 109–110, 164–165
Graham, Mark, 75
Grinding, 206n81
Grubhub, 9

Hackers, 24–25
Hall, Stuart, 42, 170, 178
Hallinan, Blake, 35
Halperin, Yoav, 146
Hammami, Nadim, 104, *105f*
Hannerz, Ulf, 186
Harindranath, Ramaswami, 187
Hashtag activism, 10, 144, 150–154
Hashtag hijacking, 10, 150–154
Heeks, Richard, 62
Heinwhoner, Rachel, 7, 22
Hesmondhalgh, David, 41
Hidden transcripts of resistance, chat groups as, 96–97
Hinton, Geoffrey, 159
Hive, 158
Hobsbawm, Eric, 175–176
Hollander, Jocelyn, 7, 22, 169
Honor of Kings (game), 68, 201n19
Horkheimer, Max, 156
Hybrid ethnography, 185–186

IG Metall, 130
Indigenous peoples, algorithmic discrimination against, 4
Individual practices of resistance, food delivery couriers, 76–85
  automatic booking of working shifts, 79, *80f*
  delivering outside platform, 84
  diaries as "technologies of the self," 77–78
  multiple accounts on same platform, 81
  order refusals, 81–82
  order stealing, 60–61, 85
  route shortcuts, 82–83
  Shuadan, 83–84
  working for multiple platforms, 79–80
InDriver, 181
Industrial capitalism, 73–74, 172
Industrial Revolution, 89, 173
Influencers. *See* Cultural industries
Information disorder, 134
Infrastructural studies, 14–15
Infrastructural tactics, chat groups as, 91–100
  hidden transcripts of resistance, 96–97
  learning environments, 93–96, 163–164
  as mutual aid and solidarity-building spaces, 97–99, 163–164
Infrastructure, impact of platformization on, 110–111
Instagram. *See also* Engagement groups
  content moderation avoidance strategies on, 146–147
  folk theories of, 95, 120
  moral economies of, 36, 41, 122–127, 168–169
  platform power of, 172
  shadow banning on, 122
  strategic algorithmic agency on, 48
  tactical algorithmic agency on, 54–55
  visibility on, 164

Instant messaging apps, ethnographic research conducted in, 182–183
Institutional/strategic algorithmic politics, 10, 132, 134–137
Instrumentarian power, 3–4, 14
Interviews, 181–183, 184, 185
I Promessi Sposi (Manzoni), 31–32
Isin, Engin, 132

Jackson, Sarah, 152
Jansen, Till, 19
Jaume I University, Spain, 181
Jenkins, Henry, 41
Jitsi, 179, 188
Johansson, Anna, 74
Jones, Phil, 158
Just Eat, 79, 87–88, 104, 181. *See also* Online food delivery platforms

Kant, Tanya, 45, 141
Kapsch, Patrick H., 42
Kaun, Anne, 18, 45, 141, 147
Keller, Tobias, 143
Kellogg, Katherine, 216n50
Kennedy, Helen, 138
Kitchin, Rob, 18
Klein, Klein, 148
Klinger, Ulrike, 143
Knights League (KL) WeChat group, 99
Koopman, Colin, 3
K-pop fan communities
  algorithmic agency of, 107–108, 115–117, 152–153
  solidarity and collectivist values in, 174–177
Kropotkin, Pyotr Alekseevic, 164

Langlois, Ganaele, 14
La Pajara ciclomensajeria, 104
Latour, Bruno, 35
Learning environments, chat groups as, 93–96, 163–164
Legal fragmentation, of food delivery couriers, 73

Lie, Merete, 34
Livingstone, Sonia, 16
Location-based services, 62
Logouts, solidarity in, 86
Lomborg, Stine, 42
López Ferrandez, Francisco Javier, 181, 202n22
LoQueSigue, 151
Lost Illusions (de Balzac), 111
Love bombing, 103
Luddism, 89–90, 175
Lupton, Deborah, 187
Lyft, 9

Macroinfluencers, 120
Madge, Clare, 186
#MAGA hashtag, 153
Magalhães, João, 150
Malaysian peasant resistance, 22, 25–26, 90, 128
Malik, Kenan, 165
Maly, Ico, 135, 139
Manriquez, Mariana, 206n81
Manzoni, Alessandro, 31–32
Mar, Mba, 173–174
Markets, shifting of, 109
Matsutake mushroom hunters, 207n83
Matthew effect, 55
Mauss, Marcel, 128
McQuillan, Dan, 140
Media, automated, 156
Media repair practices, 45, 147
Megainfluencers, 120
Meituan, 63, 68, 181. *See also* Online food delivery platforms
Mejias, Ulises, 3
Mensakas, 104–105
Meta-data optimization, 49–50
#MeToo hashtag, 144
Miceli, Milagros, 157
Microinfluencers, 120–121, 124, 211n56
Microresistance, 25
Microsoft, 13
Mighty AI, 158

Milan, Stefania, 23, 133, 139, 182
Minorities, algorithmic discrimination against, 4
Molesworth, Kat, 130
Moral economies, 7–8, 37–38. *See also* Platform moral economies; User moral economies
algorithmic agency shaped by, 40–43
algorithmic alliances and, 56–57
in algorithmic politics, 143–147
concept of, 10, 29–30
in cultural industries, 36, 41, 122–127, *126f, 127f,* 168–169
emergence of, 31–34
gaming versus optimization distinction in, 38–40
of gig workers, 100–104, 206n81
platform paternalism and, 38–40
relevance of, 165–169
Moral Economy of the Peasant, The (Scott), 33, 123
Moran, Rachel E., 147
Morris, Jeremy Wade, 18, 49, 110
Mother Cyborg, 24
Mozilla Foundation, 216n60, 224n61
Mueller, Gavin, 90
Multiple accounts of food delivery couriers, 81
Multiple platforms, couriers working for, 79–80
Multisited ethnographies, 186
Murtula, Francesca, 180, 184, 188
Musk, Elon, 159
Mutualism, 167. *See also* Collective action; Solidarity
in engagement groups, 120, 127–130
exchange of working hours, 87–88, *88f*
in private online chat groups, 93–100, 163–164, 173–175
#myNYPD hashtag, 152

Nagy, Peter, 19
Namma Yatri, 177

Nanoinfluencers, 120–121, 211n56
Narrative agency, 142
Narrative capacity, 142
Neff, Gina, 19
Negotiated decoding, 42
Negroponte, Nicholas, 223n44
Negus, Keith, 116
Neoliberalism, 112, 126, 157
Netflix, 111, 168
Neumayer, Christina, 141
New York City Police Department,
    public relations campaign by, 152
Nieborg, David, 9, 108, 110, 114, 172
Nissenbaum, Helen, 150
Noble, Safiya Umoja, 3, 140
Nucera, Diana, 24
NVivo, 185

Obama, Michelle, 147
Obfuscation, algorithmic evasion and,
    150
#OccupyWallStreet hashtag, 142
Ocran, Nicole, 130
Ojol (mobility platform drivers), 174
O'Meara, Victoria, 117–118, 120
Ong, Jonathan, 137
Online food delivery platforms, 9
  agency and resistance of couriers/
    drivers for. See food delivery
    couriers
  algorithmic governance in, 63–67,
    164–165
  computational power of, 63–67
  definition of, 61–63
  digital labor exploited by, 4–5
  dispatchers, 65–67, 66f
  emergence of, 63
  emerging alternatives to, 104–106
  history and growth of, 63
  moral economies of, 100–104
  operational logic underlying, 61–63
  platform capitalism and, 73–74
  politics of, 90–91

ranking systems and gamification,
    67–71, 69f, 72f, 80, 82, 93, 182,
    206n81
  relevance of, 161–165
  research methodology and, 180–183
Onuoha, Mimi, 24
OpenAI, 158
Oppositional decoding, 42
Oppositional play, 206n81
Oppositional solidarity, 99–100
Optimism, 10–11
Optimization
  cultural, 48–50
  gaming versus, 38–40
  strategic algorithmic agency, 48–50
  tactical algorithmic agency, 51–52
Order refusals, by food delivery couriers
  coordinated, 86–87
  individual, 81–82
Order stealing, by food delivery couriers,
    60–61, 85
Organisation for Economic Co-operation
    and Development (OECD), 62
Organizational fragmentation, of food
    delivery couriers, 73

Pais, Ivana, 62
Paternalism, platform, 39–40
Peck, Jamie, 73
Peer-to-peer (P2P) networks, 41
People's Guide to AI, A, 24
Pessimism, 10–11
Peters, John Durham, 14
Petre, Caitlin, 39
Phillips, Rachel, 73
Physical gig economy, 62
Platform capitalism, 10, 36, 73–74, 157,
    159, 172–173
Platform cooperativism, 25, 35, 104–106
Platform Cooperativism Consortium, 106
Platformization
  consequences of, 9, 14, 123, 162
  dimensions of, 9, 108–114

platform working class, making of, 171–178

Platform labor, 4–6

Platform moral economies
algorithmic agency and resistance and, 30, *31f*, 48–56, 170–171, *171f*
concept of, 34–36
relevance of, 165–169
tension with user moral economies, 38–40, 100–104

Platform paternalism, 39–40

Platform power
algorithmic bias/discrimination, 3–4, 15–16, 54, 140, 189n9
algorithmic visibility, 9, 111–114, 142–143, 164, 208n22
asymmetrical nature of, 3–4
audience/user appropriation of, 16–18
consequences of, 13–16
in cultural industries, 108–111
governance and, 109–110
infrastructural transformation resulting from, 110–111
limitations of, 16–18
market shifts and, 109
of online food delivery platforms, 61–71
pervasiveness and invisibility of, 21
platform paternalism, 39–40
platform working class, making of, 171–178

Platform studies, 14, 161–162

Platform working class, making of, 171–178

Pods. *See* Engagement groups

Poell, Thomas, 9, 39, 108, 110

Political bots. *See* Astroturfing

Politics, algorithmic, 8–9
agnosticism of, 10, 139, 154
algorithmic amplification, 10, 143–147, 154

algorithmic evasion, 147–150, *149f*, 154, 219n109
algorithmic hijacking, 150–154
algorithmic populism, 133, 135
algorithms as repertoire, 139–140
algorithms as stakes, 139–140
contentious/tactical, 10, 137–142
emergence of, 10, 131–133
institutional/strategic, 10, 132–133, 134–137
key takeaways, 153–154
of online food delivery platforms, 90–91
relevance of, 161–165
research methodology and, 184–185

Poor, algorithmic discrimination against, 4

Populism, algorithmic, 133, 135

Posada, Julian, 157

Positionality, 11–12

Post-truth, 134

Practice of Everyday Life, The (de Certeau), 47

Precarization, 43

Prey, Robert, 16

Procter, James, 170

Pro–eating disorder (Pro-ED) communities, content moderation avoidance strategies of, 146–150

Propaganda, computational, 10, 41, 46, 53, 133, 135, *171f*

Protest, impact of algorithms on, 10

Protests, 71–73, 88–90

Public transcripts of resistance, 96

Putri, Lamia, 116

Qadri, Rida, 91, 173

Quandt, Thorsten, 134

Randall, Adrian, 33, 104

Ranking systems, of online food delivery platforms, *67–71, 69f, 72f*

Rappi, 181

Raval, Noopur, 173
Reflexivity, agency and, 19
Repertoire, algorithms as, 139–140
Repertoire, data as, 23, 139
Repertoires of contention, 10, 141–142
Reputational wars, 54
Research methodology, 11–12
  conversations and planning, 179–180
  cultural industries, 184
  desk research and other data, 188
  ethnographic research, 185–186
  food delivery platforms, 180–183
  methodological imagination, 180
  politics, 184–185
  screenshots, power of, 187–188,
    226n28
WhatsApp's role in, 182–184, 186,
  187–188, 226n28
Resistance. *See* Algorithmic resistance
Resistance through Rituals (Hall),
  170
Restaurant-led *Shuadan*, 83–84
Review boosting, Tripadvisor, 54
Review vandalism, 54
Ride-hailing platforms, 62–63, 71, 91
Riots, 71–73, 88–90
#RipInstagram hashtag, 117
Ritual resistance, 170
Robin Food, 104
Rodant, 104
Round pods, 120
Route shortcuts, by food delivery
  couriers, 82–83
Roy, Srirupa, 136
Ruppert, Evelyn, 132
Russell, Stuart, 155

Sabotage, 88–90
Sadowski, Jathan, 16
Sajaegi, 116–117
Sayer, Andrew, 33
Scale, 158
Scholz, Trebor, 6, 106

Scott, James, 7, 22, 25–26, 33, 58, 74,
  90, 96, 123, 128, 163, 169, 175, 177,
  194n49, 195n1
Seaver, Nick, 167
Self-driving vehicles, 158
Service Platform Couriers (SPCs), 68,
  201n20
SF Express, 181
Shadow banning, 35, 122
Shifman, Limor, 34
Shifting markets, platformization and,
  109
Shuadan, 83–84
Signal groups, 11, 143–144
Siles, Ignacio, 42, 187
Silverstone, Roger, 26
Simmel, George, 21
Sin Delantal, 181
Singh, Swati, 61, 181, 200n3
Skype, 188
Smith, Adam, 33
Social alliances, 57
Social fragmentation, of food delivery
  couriers, 73
Social movements, impact of algorithms
  on, 10
Social Sciences and Humanities Research
  Council of Canada, 185
Sociomaterial alliances, 56–57
Sock puppets, 135
Solidarity. *See also* Collective action
  engagement groups, 117–130, *125f,
    126f, 127f*
  entrepreneurial, 37, 99–100
  oppositional, 99–100
  private chat groups, 97–99, 163–164,
    173–175
  solidarity logouts, 86
Sonic optimization, 49–50
Soriano, Cheryll Ruth, 37, 100, 129
Spasmodic view of popular history, 32
Spatial fragmentation, food delivery
  couriers, 73

Spoofing, 19, 107

Spotify
algorithmic agency and resistance to, 26, 49–50, 107–108, 115–117
algorithmic alliances with, 56
algorithmic governance of visibility in, 164, 222n31
moral economies of, 41–42, 168–169
platformization and, 109–111
platform power of, 172

Sprave, Jörg, 130

Stakes, algorithms as, 139–140

Stakes, data as, 23, 139

Stakhanov, Alexei, 206n82

Stakhanovite, 206n82

Stark, David, 62

Stealing of orders, by food delivery couriers, 60–61, 85

Stiegler, Bernard, 156

Strategic algorithmic agency, 8
in cultural industries, 116–117
moral economies and, 30, *31f*, 48–54, 170–171, *171f*
in politics, 132–133
tactical algorithmic agency versus, 43–47

Stream parties, 107

Strikes, 71–73, 88–90

Structuration theory, 7, 20

Stuart, 70–71, 181. *See also* Online food delivery platforms

Sumseuming (streaming 24/7), 107–108, 115–117

Surge clubs, 9

Surge pricing, 71

Surveillance capitalism, 14, 17, 58, 81, 156

Švelch, Jan, 187

Swiggy, 59–61, 63, 97, 181. *See also* Online food delivery platforms

Symbiotic agency, 19–20

SyRI (system risk indication) system, 189n9

Tactical agency, 141

Tactical algorithmic agency, 8
in cultural industries. *See* cultural industries
of food delivery couriers. *See* food delivery couriers
moral economies and, 30, *31f*, 51–56, 170–171, *171f*
in politics. *See* politics, algorithmic
strategic agency versus, 43–47

Tarrow, Sidney, 133

Tax fraud, SyRI (system risk indication) system, 189n9

Technological fragmentation, of food delivery couriers, 73

Telegram groups, 11, 91–92. *See also* Engagement groups
algorithmic activism on, 143–144
coordinated order refusal through, 86–87
folk theories of, 95, 120
as hidden transcripts of resistance, 96–97
as learning environment, 93–96, 163–164
as mutual aid and solidarity-building spaces, 97–99, 163–164, 173
research methodology and, 182–184, 186

Tencent Games, 13, 201n19

Terkalas, Shayla, 211n63

Terms of service (ToS), 6, 29, 124

Terranova, Tiziana, 6

Tesla, 158

Thompson, Edward P., 32–34, 40, 58, 104, 171–173, 174, 195n1

TikTok. *See also* Engagement groups
moral economies of, 124
solidarity on, 99
visibility on, 164

Tilly, Charles, 133, 140–141

TiMi Studio Group, 201n19

Tinder
  algorithmic agency on, 19, 26,
    48–52, 55
  research methodology and, 188
Tønnesen, Ozlem Demirkol, 149
TOR, 150
Trade unions, 80, 90, 172, 174, 202n23
Transport automation industry, 158
Tripadvisor, review boosting on, 54
Troll farms, 53–54, 136–137
Trump, Donald, 35, 115, 134, 146–147,
  152
Tsing, Anne Lowenthal, 207n83
Tufekci, Zeynep, 142
Tumblr, 146–150
Tuyul apps, 91
Twitter, 146–150
  algorithmic activism on, 143–144
  algorithmic alliances, 56
  algorithmic evasion on, 147–150
  algorithmic governance of visibility
    in, 164
  algorithmic hijacking on, 150–153, 154
  content moderation avoidance
    strategies on, 146–147
  moral economies of, 36, 168–169
  platformization and, 109
  platform power of, 172
  research methodology and, 185

Uber, 181
  algorithmic agency and resistance to,
    9, 19, 54–55, 103
  Mighty AI acquired by, 158
  moral economies of, 36, 41, 206n81
  platform labor exploited by, 1, 5
Uber Eats, 181. *See also* Online food
  delivery platforms
  algorithmic agency and resistance to,
    59–61
  history of, 63
  ranking systems and gamification,
    70–71, *72f*

Unione Generale del Lavoro (UGL),
  202n23
Unions
  cultural industry, 129–130, 177
  food delivery courier, 92, 102
  trade, 80, 90, 172, 174, 202n23
University of Greenwich, 181
University of Guanajuato, Mexico, 181
University of Siena, 184, 188
Upwork, 5, 62
User moral economies
  algorithmic agency and resistance
    and, 30, *31f*, 170–171, *171f*
  concept of, 36–38
  oppositional, 37
  relevance of, 165–169
  tension with platform moral econo-
    mies, 38–40
Utilitarianism, in engagement groups,
  127–130

Vaccine-opposed groups, content
  moderation avoidance strategies of,
  146–150
Valentine, Melissa, 216n50
Vallas, Steven, 21, 22
van Dijck, José, 35, 108
van Doorn, Niels, 67
Vasudevan, Krishnan, 206n81
Velkova, Julia, 18, 45, 141, 147
Villeneuve, Denis, 159
Vinted, 124
Vinthagen, Stellan, 74
Virtual private networks (VPNs), 107,
  115–116
Visibility, 9, 111–114, 142–143, 164,
  208n22

Wajcman, Judy, 168
WeChat, 86, 91–92
  as hidden transcript of resistance,
    96–97
  as learning environment, 93–96

as mutual aid and solidarity-building
    space, 97–99
research methodology and, 181,
    182–184
Welles, Brooke Foucault, 152
WhatsApp, 91–92. *See also* Engagement
    groups
    algorithmic activism on, 143–144
    algorithmic agency and resistance
        with, 11, 45, 86, 141, 226n28
    coordinated order refusal through,
        86–87
    as hidden transcripts of resistance,
        96–97
    as learning environments, 93–96,
        163–164
    as mutual aid and solidarity-building
        spaces, 97–99, 163–164, 173
    research methodology and, 179–188
#whitelivesmatter hashtag, 153
Wikileaks, 134
Winner, Langdon, 34, 90, 168
Wire, The (TV series), 165
Women, algorithmic discrimination
    against, 4
Woodcock, Jamie, 94
Workerism, 176

#YaMeCanse hashtag, 151–152
#YoSoy132 hashtag, 142
Young people, algorithmic discrimina-
    tion against, 4
YouTube, 99
    algorithmic agency and resistance
        with, 60, *60f*, 117
    platformization and, 109
    platform power of, 172
    visibility on, 164
YouTubers Union (YTU), 130
Yu, Zizheng, 181, 202n21

Zampate Zaragoza, 104
Zello, 98

Zhang, Qian, 116
Ziewitz, Malte, 40, 43
Zomato, 60, 63, 68, 181. *See also* Online
    food delivery platforms
Zoom, 179, 188
Zuboff, Shoshana, 3, 14, 16, 58, 176
Zuckerberg, Mark, 160